Computers in Amateur Radio

Third edition

Edited by Lorna Smart, 2E0POI

Radio Society of Great Britain

Published by the Radio Society of Great Britain, 3 Abbey Court, Fraser Road, Priory Business Park, Bedford MK44 3WH. Tel: 01234 832700. Web: www.rsgb.org

First published 2010
Second edition 2012
Third edition 2022

ISBN: 9781 9139 9528 7

Cover design: Kevin Williams, M6CYB

Production: Mark Allgar, M1MPA

Typography and design: Mark Pressland

Printed in Great Britain by CPI Antony Rowe of Chippenham, Wiltshire

Any amendments or updates to this book can be found at: www.rsgb.org/booksextra

Contents

A 'Thank You' to our Authors and Collaborators

This book has been updated from the revised edition in 2012. All our authors and those working behind the scenes have put so much time and effort into what they have produced. The wide variety of knowledge and experience has ensured that the chapters are as full of relevant, up to date information and advice as they possibly can in such a rapidly evolving area of technology.

Thank you very much to all of you past and present who have worked so hard to make this book a useful resource for the amateur radio community to learn more about the role that computers can have in the hobby. Some parts of this book have been reused from previous versions where the information is still relevant.

Lorna Smart, 2E0POI

1

Introduction

by Sean Gilbert, G4UCJ

Computers are no longer the equipment of the rich. Mass production and the popularity of the computer have helped to drive down the prices. Changes in technology and the forms a computer can take these days have also made them more available to people. Their functionality has also expanded with major developments in areas such as software, memory capacity, programming capabilities and portability.

Radio amateurs were quick to discover that a computer could be used in conjunction with their hobby to make operating easier or provide something extra at the station. In 1981 IBM developed what was to become known as the 'Personal Computer' or PC, this was designed for businesses and was initially beyond the means of all but the wealthy. The general public were being introduced to using computers in their homes and schools by such manufacturers as Sinclair, Commodore and Amstrad. The BBC 'Micro' was widely used in schools during the early 1980s.

Computers are now used in the radio shack for tasks such as electronic QSO logging, transceiver control, running Software Defined Radio (SDR) receivers, datamode decoders, etc. In the 21st century, just about everybody has some computing power at their disposal, be it a mobile phone, television, games console, etc. The birth of the internet proved to not only be a world-wide game changer, it also became a very useful tool for the radio amateur. The personal computer, together with internet access, is now part and parcel of many shacks around the world.

For the purposes of this book, the computers discussed will primarily be the PC which will usually be running a version of the Microsoft Windows operating system. Desktop PC's are easily modified and upgraded, with the various connectors and internal components being of a standard type. Laptop computers can sometimes be upgraded but not in as many ways. It is usually only with extra RAM or a larger/solid state (SSD) hard-drive.

Selecting a Computer for The Shack

When selecting a computer to use in the shack, certain criteria needs to be considered. The computing power of the average modern mid-range pc should be perfectly adequate for most purposes. A lower-end computer may struggle to keep up with the demands put upon it should several CPU intensive tasks be attempted at the same time. If you are fortunate enough to have a computer to use just for radio and sufficient room, using multiple monitors becomes a realistic possibility. A second (or third) monitor is a useful addition to the shack. If space is at a premium, there are some very small but powerful computers available. These computers can even be in the form of a USB dongle that plugs into an existing screen (usually a TV) and offer reasonable performance. Due to lack of physical space, they tend to have a reduced number of in/out ports and few ways to upgrade the hardware.

Audible Noise

The last thing you are likely to want in your shack is a computer that has fans so noisy they can be heard when you transmit, or disturb family members if the computer is in a shared location, so select a machine with quiet fans. CPU fans tend to be the noisiest, but the type of fan used depends on the heatsink on the CPU chip. A larger, more efficient heatsink enables the use of a slower, quieter fan. There has been a shift towards larger diameter CPU fans (120mm is not uncommon), mounted on more efficient heatsinks that offer better heat exchange properties. The larger fans spin at a slower rate but draw/blow a larger amount of air through the heatsink, achieving better efficiency, with less noise. Some higher-end machines use water-cooling systems. Having water inside a PC sounds counter-intuitive, but liquid cooling methods offer far greater efficiency than conventional air-cooled systems. Laptops do not, as a rule, have fans on their power supplies and their CPU fans are, usually, fairly quiet. With all types of computers, airborne dust can block air vents, clog up heatsink fins and fans if they are not cleaned regularly. A dusty computer or laptop will run louder and hotter than a clean, dust-free system. With hard flooring, more loose dust is sucked into the case, building up very quickly. In extreme cases, dust can clog the fan and heatsink to such an extent the computer overheats and stops working altogether.

Electrical Noise

Some desktop computer cases are of a better quality and offer better screening than others. Equally, more expensive power supplies are likely to incorporate better filtering. Generally speaking, the more you pay for a case, the better screened it is likely to be; and the more you pay for a power supply, the better filtered it is likely to be. Laptop computers employ external power supply modules and some of these are more electrically noisy than others. Users should be particularly wary of cheap, often inferior quality, third party replacements as they can interfere with vast swathes of the HF spectrum, and beyond. If you are going to buy a new computer from a shop, it can be useful to take a portable receiver with you and check how much noise emerges from the case, cables, power supply etc., by placing it close to each of them in turn. You might get some puzzled looks from the shop staff, but you won't be laughing if you get your new machine home and discover that it causes interference. The EMC chapter of this book goes into this in more depth.

Connectivity

First and foremost, make sure your computer has all the ports that you are ever likely to need. Few modern PCs have some of the older ports such as parallel printer ports (sometimes very useful for keying). If required, you can use plug-in cards to provide you with various interfaces.

Most computers have a selection of USB type 'A' ports, If the socket has a white or black plastic tab, the port will be USB 2. If the tab is blue, the port is a USB 3.x type. Other colours, such as red or yellow indicate high current or fast charging sockets. USB ports are backwards compatible, but not all devices will function correctly if used with a more modern USB port. Likewise, some high bandwidth/high current USB2 devices, such as SDR's, etc. may run more smoothly if plugged into a USB 3 socket.

If the computer does not have enough USB sockets a plug-in card or an external multi-port hub can be added economically. It is highly recommended that both external USB and internal PCI/PCIe hubs are powered and not reliant solely on the motherboard USB bus to provide adequate current. USB2 ports support a maximum current draw of 500mA per port. When an unpowered USB hub reaches full capacity, the devices may stop working. Also, sharing available bandwidth may become an issue. USB 3.x has higher current and bandwidth limits, but a powered hub is still recommended. Laptops tend to have fewer USB ports than their desktop counterparts, so the use of an external hub will likely become necessary which, again should have an external power source.

Operating Systems

Without an Operating System (OS) a computer can do nothing. The most commonly known OS in the earlier days of PC's was Microsoft DOS (Disk Operating System). DOS was a clunky command line system, requiring each command to be input via a keyboard. Early versions of Microsoft Windows, such as Windows 3.11, which were not really operating systems, they were a Graphical User Interface (GUI) to access previously written and compiled programs or procedures. For the first time, a user could move a mouse around on a screen that emulated a 'desk top' and click a picture (icon) to launch programs or files instead of having to learn several command line text strings to launch, close or configure DOS programs or writing batch files to automate the execution of multiple line commands. Windows and Linux are the operating systems most commonly found on modern computers, with Windows having by far the lion's share of installations. Below are brief descriptions of the most commonly encountered operating systems.

Microsoft Windows is, by far, the most popular operating system used by the public. To run most of the available radio software, newer versions of Windows such as 7, 8 or 10 are required. Some software will run under earlier versions such as XP or Vista, but updating to a newer version is recommended. Later versions of Windows are more secure and use resources in a more efficient manner. Windows 11 has very recently been released to the public, but is not in wide enough circulation to know what existing software will run correctly. One would hope that any software that works under Windows 10 would continue to do so under the new OS, but time will tell.

Macintosh computers (produced by Apple Inc.) are not as popular in amateur radio circles as the PC, due to the lack of radio related software for the Mac operating system (which was known, until 2016, as 'OS-X' and from then on as 'macOS').

Linux The Linux OS is, unlike the other systems, open source. This means the source code is available and can be modified and compiled by an end user to fulfil a specific need (unlike Unix, which has a proprietary, therefore protected, source code). Linux has many main versions, or distributions, as they are known. Some of the more familiar are 'Unbuntu', 'Fedora' and 'Debian' which are free to install and use, whilst 'Red Hat' is a commercial distribution.

Raspberry Pi mini computers work with many of the Linux operating systems. One of which is called 'Raspberry Pi OS' (previously known as 'Raspbian'). The Raspberry Pi (RPi) is becoming more and more popular with radio amateurs due to their compactness as well as being surprisingly powerful and inexpensive, when compared to a full size PC.

It is possible to install an RPi in a remote location to run things as diverse as weather stations, ADS-B virtual radar systems (for tracking aircraft beacon broadcasts) or hosting a web SDR. Being self-contained and remotely accessed, the RPi installation can be in an RF quiet location and does not have to be near to the user. In isolated positions, power could be derived from solar panels, etc. and use a 4G dongle for internet connectivity. This is becoming a more and more appealing idea amongst the amateur radio and SWL community due to the increasingly high levels of RF pollution that is now to be found in most urban locations.

Performance

As computers are put to ever increasing use in the radio shack, one area that must not be overlooked is how well it performs the required tasks. Computers have a finite set of resources which must be divided amongst all operations that take place whilst running the operating system and any additional software. The two easiest ways computer performance can be improved are by: 1) Increasing the amount of RAM (Random Access Memory) installed and 2) Exchanging the mechanical hard disk drive (HDD) for a solid state drive (SSD). 8GB is enough RAM to cope with most tasks, but particularly resource heavy software may need 16GB, or more, to avoid the system becoming unresponsive.

Software Packages

Software (programs) familiar to most are those used for e-mail, internet access, word processing, spreadsheets, drawing, digital photograph manipulation and storage, etc. There are numerous specifically written programs that are of real use to radio amateurs. In a number of subsequent chapters, contributing authors take a detailed look at some of those software packages.

Software, whether for the radio amateur, or not, develops over time. Bugs are fixed, functionality improved, the appearance may change, and so on. It is important to keep the operating system up-to-date, with any additional security patches installed. Likewise, software should be kept current. Before installing an updated version of any software, ensure existing data is backed up (preferably to an external media device) so it may be restored should the update cause issues or the original become corrupted. Checking the internet should reveal the latest available versions of the software.

Where possible, install a trial version of the required software, to ensure system compatibility, before purchase. When downloading a program or software package, be certain the site is genuine. Most websites now have the prefix 'https://', indicating the site is secure, but vigilance is still required as cyber criminals become increasingly clever at disguising their activities. Before running any software, perform a scan using antivirus software.

Search engines are a very handy tool when looking for software. I suggest trying a number of search engines as some may list sites that are not shown on others. The hobby of amateur radio is extremely well served when it comes to software packages, because a lot of radio amateurs are technically-minded and have an interest in programming. Those who are capable of doing so tend to develop programs to address particular needs, and are often happy to share the solution with a wider audience. We have our fellow, philanthropic radio amateurs to thank for the many programs designed for our hobby that are free to use. Some software, particularly the more complex and time consuming variety, may require the payment of a fee, after an initial trial period has expired.

2
Data Modes

by Andrew Barron ZL3DW

Amateur radio data modes carry digital data, usually text, or the contents of computer files, which could include pictures, data, or programs. The term has fallen out of use now, and most operators refer to 'data modes' as being 'digital modes.' Which is a wider term that includes communication systems which encode voice signals for digital transmission, such as DMR, D-Star, and P25, as well as 'picture' modes like SSTV.

Digital mode transmission refers to any communication over radio where information is converted into a digital signal before being transmitted and converted back to the original format at the receiving station. On the HF bands, it is usually achieved by modulating our SSB transmitters with changing or multiple tones. This is known as frequency shift keying, (FSK). We also use amplitude shift keying (ASK) in the case of CW (Morse Code) operation and Hellschreiber, and phase shift keying for PSK-31 and QPSK operation. On the VHF and higher bands, digital voice modes are transmitted using switched audio tones (FSK) or phase shift keying (PSK) of FM transmitters.

Packet Radio and APRS also use FSK on FM transmitters. CW is used occasionally for DX operation and on amateur radio satellites. The WSJT modes including JT65, FT8, and WSPR use SSB transmitters, for EME (Earth-Moon-Earth), meteor scatter, Troposcatter, propagation reporting, and other long-range DX contacts.

The Computer Connection

Computers are essential for the data modes. A few of the newer radios have PSK-31 and/ or RTTY encoders and decoders built-in, but for the majority of users and modes, a connection between your radio and a personal computer (PC) is necessary. 'Digital mode' software on the PC creates the audio signal, or keying in the case of CW and FSK (frequency shift keying) modes, for the transmitter, and decodes and displays the text or data from the received audio. You usually get some sort of instant message macros and call logging. New radios use a single USB cable to carry the transmitter and receiver audio, CW or RTTY keying, and the PTT (press to talk) signal which makes the transceiver shift from receiving to transmitting.

Older radios use the PC soundcard for the audio connections and RS-232 serial port (RTS and DTR) lines for the keying signals. Having a direct connection between your radio and some older computers created an earth loop, where 50Hz or 60Hz hum affected the audio. Earth loops can occur when there are two slightly different length earth connections between devices. For example, the earthed side of your audio leads and the 'mains' earth through the power supply on the PC and the DC power supply for the radio. Hum is capacitively or inductively coupled in and travels around the loop. This problem was solved by employing an interface box such as a "Rig-blaster" which had optocouplers to isolate the RS-232 switching signals and transformers to isolate the audio input and output. Using a single USB cable is tidier and there is no need for interface boxes. The USB driver software for the radio creates a software CODEC (coder-decoder) that "looks" like an audio soundcard to the computer operating system and the digital mode software. You select the CODEC in the program settings, and adjust the audio levels, via the digital mode software, the Windows sound settings, and/or the menu settings in the radio. The USB driver also creates one or two 'virtual' COM ports. These appear to the computer to be the same as a hardware RS-232 serial port on the radio, but usually only the DTR (device terminal ready) and RTS (ready to send) RS232 signalling lines are used. You should take a note of the COM port number(s) that are created by the USB driver software and set the data speed the same on the PC program and the radio. Set parity to N and bits to 8. 'Handshaking' is used by some software, but it can usually be set to 'Off' or 'Hardware.'

I normally use RTS (ready to send) for the transceiver PTT control. Because when you have enabled the PTT, you are "ready to send" the data. That leaves the DTR line for sending CW or FSK RTTY.

Once you have the USB driver software installed and the USB cable connected, you can set the COM port settings on the radio and the PC and set the audio levels. Use the minimum amount of audio into the transmitter that will result in full RF output power when the mode is transmitting data. More input signal will result in the transmitter applying ALC (automatic levelling control) and this can compress the data signal. As an initial guide, set the level from the radio into the computer at about 50% of the scale measured on the digital mode software. These settings should only need to be made once. After that, you will be able to try out any of the various data modes available in the digital mode software.

History

Several of the older data modes pre-date computers, and a few of them pre-date radio.

Morse Code (not the currently used 'International' Morse code) was used on telephone wires from around 1837. By 1844 'Telegraphy' machines were able to mark a paper tape with the Morse characters, so the code could be read sometime after it was received. As soon as the 'Spark-gap' transmitters were invented in the 1890s, Morse Code was employed to send and receive "wireless" messages. It revolutionised communication with ships. Before radio, the only way to pass 'ship to ship,' or 'ship to shore' messages was 'visual line of site' signalling systems, such as semaphore, lights, and flags. Radios started to be used in Zeppelins, airships, and aeroplanes during World War One. Again, all transmissions were by CW. At the time, AM radio transmitters were too large and heavy to be carried by aircraft, although the US Navy published a 'white paper' outlining a plan to use AM radio to communicate with aeroplanes as early as July 1917.

CW

Many CW purists don't agree that CW (using Morse code) is a data mode. But it uses a code to represent text characters and it transmits full RF power ASK (amplitude shift keying). The fact that you can send and decode a CW signal with a radio and a computer proves that it is a data mode. CW is unique in that it is the only data mode that can be sent and received by a human without any mechanical or electronic encoder and decoder.

Hellschreiber can be read on a waterfall display or paper tape without a decoder, but it can't be coded without a machine interface to scan the characters. Morse code is cleverly designed so that the most used characters contain the fewest dots and dashes. This speeds up the achievable transmission rate, measured in 'words per minute.' Most operators send at 12 - 25 wpm (words per minute), although faster speeds are common during contests. Few operators can read or send Morse by hand at greater than 60 wpm, but some computer software can work at more than 100 wpm. Very low-speed CW beacons are used on the LF (low frequency) bands to achieve communication over very weak signal paths.

Keying an SSB transceiver with a 25 Hz

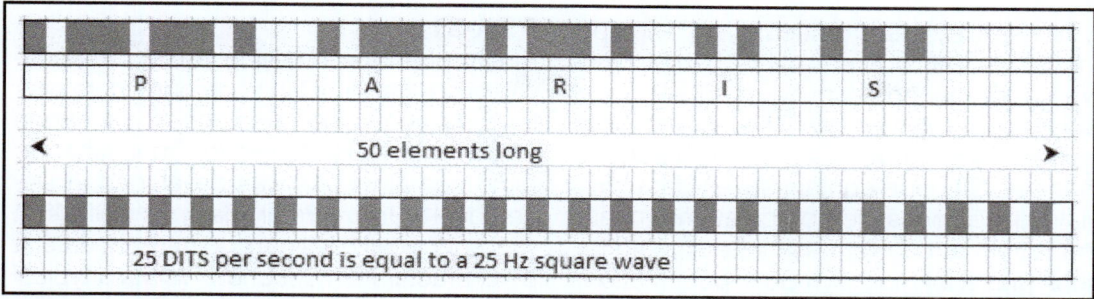

Fig 2.1: Keying an SSB transmitter with a 25 Hz square wave is equivalent to sending CW at 60 wpm.

square wave is equivalent to sending CW at 60 wpm. The word, 'PARIS' has been selected as a typical average-sized word for CW speed testing.

Sending PARIS plus a word space in one se cond is recognized as sending at a rate of one word per second, or 60 words per minute. The 'dits,' 'dahs,' and spaces of PARIS plus a word space are exactly 50 elements long. A string of 25 'dits' is also 50 elements long. If a string of 25 'dits' is sent in one second, it is being sent at a frequency of 25 Hz.

RTTY

RTTY (radio teletype) evolved from teletype machines (teleprinters) which transmitted the 5-bit Baudot code to send text and numerals over telephone lines and other cables. These machines were still being used to send Telegrams, business-to-business information, stock market data, and weather messages until the early 1980s when fax machines and computer networking via audio modems on the phone lines, and eventually email via the Internet, made them redundant.

The US Navy successfully tested RTTY communications between an aeroplane and the ground in 1922. After WWII, amateur radio operators started to buy surplus teleprinters and use them on the amateur radio bands. Initially using ASK, because the FCC did not approve frequency shift keying until 1951.

The Baudot code supports upper case characters, numerals, and a few punctuation marks. Amateur radio RTTY signals usually use a 170 Hz shift and transmit at 45.45 bauds (about 60 wpm). It requires a receiver bandwidth of around 250 Hz. More than double the bandwidth required by PSK31. Commercial RTTY is often at 50 bauds and uses different tone spacings, often 200 Hz. Because the transmission is always one tone at a time (Mark or Space), you can use a more efficient, non-linear, Class C RF amplifier. It is commonly created in one of two ways. Either a keying signal is used to shift the transmitter frequency up by 170 Hz when a Space signal is to be transmitted (FSK mode), or the digital mode software creates two audio tones one for the Space signal and one for the Mark signal. Typically, RTTY is transmitted on the lower sideband (LSB) with, Mark = 2125 Hz and Space = 2295 Hz. This audio method using two tones is called audio frequency shift keying (AFSK). CW can be sent the same way, using a single audio tone at the same frequency as the CW receiver offset (typically 600-700 Hz). The MFSK modes are created the same way using multiple audio tones. (More about those later).

Even though the mode is quite slow and prone to decoding errors, RTTY is still popular and there are several international contests and awards devoted to it.

ASCII and Unicode

The original ASCII (American Standard Code for Information Exchange) code was used by electromechanical teletype machines that worked at 110 bauds. This was a significant improvement over the old Baudot teletype machines which only worked at 45.45 bauds. Also, the characters in the ASCII code contain seven data bits. Having 128 possible combinations meant that upper and lower case letters could be sent, plus a lot more symbols, punctuation, and control characters.

As standard, ASCII does not incorporate error detection or correction, but it is possible to add a measure of error detection by adding a parity bit, making an 8-bit ASCII code.

ASCII is not usually sent directly in the way that RTTY is used to send the Baudot code, but most text-based data modes send standard ASCII characters along with other synchronisation and often forward error correction bits (FEC).

The problem with the ASCII character set is that there is no standard format. The code was modified to fit the requirements of different languages. For example, the European version has UK pound and Euro symbols that don't exist in the US version. It also contains accented symbols. There are Chinese / Japanese versions that contain special symbols used in those languages. Since 2010 the Unicode Consortium has been managing a standard character set to integrate all the options.

The Unicode character set also includes the Emojis used on your phone. Unicode uses several UTF (Unicode transformation standards) that use as many as 32 bits to carry the 1,112,064 valid character codes of Unicode. UTF-8 uses up to four 8-bit bytes per character. Microsoft Windows

UTF-32 is a fixed-length format that always uses 32-bits (4 bytes) per character. Any data mode that can transmit pure digital data such as computer files can transmit Unicode characters, but most modes are concerned with transmitting text as efficiently and quickly as possible, so the old 7-bit (or 8-bit with parity) ASCII characters are used.

PSK-31

Binary phase shift keying (BPSK) was developed for transmitting computer data over long distances using cables or telephone wires. The 'PSK-31' version that amateur radio operators use was developed in December 1998 by Peter Martinez G3PLX. It was specifically designed to transmit data at 31 characters per second which is a typical typing speed. Unlike the 5-bit Baudot code used for RTTY, PSK transmits 8-bit ASCII characters, so it can transmit uppercase and lowercase letters, numerals, and many punctuation and special characters. Baudot can only transmit a few punctuation marks, numerals 0-9, and the 26 uppercase characters.

Because of its narrow bandwidth, PSK-31 works very well in low signal conditions, coping well with path fading and multi-path fluctuations. It became very popular until the decline in sunspots at the end of solar cycle 24 made worldwide HF communications difficult. The new FT8 digital mode offers amazing performance when propagation is very weak, and it has now become the dominant digital mode for text transmissions. I hope that as band conditions improve during solar cycle 25, we will see a return to "conversation" digital modes like PSK-31 and RTTY.

Packet Radio

Packet radio was very popular before the Internet and emails made instant communication possible. It was used on the HF bands at 300 bps (bits per second) AFSK and on the VHF and higher bands at 1200 or 9600 FSK. You required a sort of modem box called a TNC (terminal node controller) to create the AX.25 format signal and to decode the received signal. The terminal was a "dumb terminal" usually a teletype terminal or an electronic terminal such as used for communicating with main-frame and mini-computers like the PDP-8 and PDP11. You could communicate radio to radio or leave a message on a 'packet server' which could be recovered by the recipient later. A bit like an email today. Some amateur radio satellites were digipeaters with this "store and forward" capability. You could upload a message to the satellite and a friend in another part of the world could download the message when the satellite passed over their location. At the time this was amazing! You could send a message or even a file or a picture, to a radio ham anywhere in the world. Before emails, this was impossible to do for free. You had to send an expensive telegram or write a letter. These days digital mode software on your PC can emulate the functions of the TNC and packet radio is just another digital mode.

"AX.25 (Amateur X.25) is a data link layer protocol originally derived from layer 2 of the X.25 protocol suite and designed for use by amateur radio operators." [Wikipedia]. What this means is that the data that is to be transmitted is arranged into data packets that comply with the AX.25 structure. It includes a data frame structure that allows the data to be recovered accurately and error checking. Once the data is in the AX.25 frames it can be transmitted using any data mode that is fast enough to keep up. In the case of Packet radio, this was 300 bauds on HF (SSB) and 100 or 9600 bauds (FM) on VHF and higher bands.

APRS

Packet radio is dead, but the AX.25 format is used for APRS (automatic packet reporting system) beacons. APRS transmits a station's position, usually derived from a GPS receiver, and a small amount of other information such as weather station data and your callsign. It has been developed since the late 1980s by Bob Bruninga WB4APR, currently a senior research engineer at the United States Naval Academy. The initialism "APRS" was derived from his call sign.

Typically, an APRS data 'beacon' is transmitted to a digipeater on a simplex VHF frequency. After the 'beacon' is received, it is repeated to other local APRS stations by the digipeater. This transmission may be picked up by area relay stations and repeated again over a wider area. One of the relay stations in the chain acts as an IGate (Internet-Gateway) which forwards the beacon data to the APRS Internet System (APRS-IS) via an Internet connection. The position information from APRS beacons all over the world is displayed on a zoomable map at https://aprs.fi. Some amateur radio satellites can 'store and forward' APRS beacons to a receiving station that is also an IGate. This allows satellite operators to send APRS beacons to the network via satellite instead of transmitting to an APRS digipeater.

AmTOR

Amateur Teleprinting Over Radio (AmTOR) was developed in 1979 by Peter Martinez, G3PLX as an improvement to RTTY. It is an adaptation of the commercial system Simplex Teleprinting Over Radio (SiTOR), used mainly by maritime stations.

AmTOR uses frequency shift keying and is sent at 100 baud, usually with a frequency shift of 170Hz. It employs a special 7-bit adaptation of the Baudot telegraphic code that contains a fixed ratio of 4 'mark' bits to 3 'space' bits. There were two AmTOR modes which both provided a level of error correction. Mode-A was a synchronised mode used when two stations were communicating. The transmitting station would send three characters. The receiving station would check the mark-space ratio and return an ACK (acknowledge signal) or a NACK (not-acknowledged) signal. When the transmitting station received the ACK signal it sent the next three characters. If it received a NACK signal it would resend the previous three characters. The synchronised mode does not work if there are two or more receiving stations such as the situation when you send a CQ message or a bulletin. In Mode-B the transmitting station simply sends each character twice. If the receiving station gets the same decode and the correct 3:4 mark-space ratio both times it prints the received character. If not, it prints an error mark.

AmTOR was usually used for point to point

links rather than general amateur radio conversations and it was very good. But the advent of personal computers has allowed for much better error correction techniques and AMTOR is very rarely used these days.

PacTOR

PacTOR was developed in 1991 by Hans-Peter Helfert, DL6MAA and Ulrich Strate, KF4KV. The basic structure is the same as AmTOR, with its fixed interval data blocks and corresponding acknowledgements, but PacTOR also combines important characteristics from Packet Radio. It uses ASCII code rather than the 7-bit modified Baudot code, so the character set is much larger. The baud rate can be automatically reduced from 200 bauds to 100 bauds, which enables traffic to move faster when propagation paths are good but still maintain contact when paths are poor. It is an FSK mode with a shift of 200Hz. PacTOR is usually only used on the HF bands. It has better error correction than AmTOR. The PacTOR II variant uses 4 phase PSK and is capable of much higher data speeds. You can still buy TNCs for PacTOR, but some digital mode programs can send and receive PacTOR without a hardware TNC.

Symbol Rate and Bandwidth

When you transmit a digital signal, each amplitude, phase, or frequency change is responsible for sending a part of the binary data stream. PSK-31 transmissions have two possible phase states. A binary 1 is indicated by a 0° phase change and a binary

0 is indicated by a 180° phase change. Each bit in the data stream causes one of those two states to be transmitted. The 32 ms period of transmission is called a symbol. Each symbol carries one bit of the data stream being transmitted.

The symbol rate of a data radio system is equivalent to the Baud rate, a term more commonly used when referring to cable transmission systems. It is the number of digital symbols sent each second.

The symbol (Baud) rate = bits per second/

bits per symbol.

The bit rate is the number of binary bits being sent per second.

Amateur radio RTTY uses two tone frequencies that are usually 170 Hz apart, or the RTTY keying moves the transmitter frequency by 170 Hz.

Commercial RTTY may use different a frequency split and may work at a higher baud rate. The code that is being transmitted is the 5-bit Baudot code rather than the 7-bit or 8-bit ASCII code that the more modern modes transmit, but the theory is the same. There are two frequency states, known as Mark and Space, so there is one symbol being transmitted for each data bit.

QPSK has four possible phase states. The data bits cause the carrier signal to change by 0°, 90°, 180° or 270°. For example, in differentially coded QPSK, an input signal of 00 causes no phase change (0°). An input signal of 01 causes the signal to change to the current phase plus 90°. An input signal of 11 causes the signal to change to the current phase plus 180°. An input signal of 10 causes the signal to change to the current phase minus 90°. Each change of phase state, symbol, is transmitting two bits of data. Therefore, the data rate is twice the symbol rate. Technically QPSK can transmit twice as much data in the same amount of time. The usual amateur radio QPSK-31 mode actually transmits data at about the same speed as PSK-31, due to the inclusion of forward error correction bits into the data stream.

A four-phase state, QPSK signal, is more difficult to decode than a two-state PSK signal. You need 3 dB more received signal to achieve the same bit error rate performance. That means the forward error correction must provide at least a 3dB improvement under weak signal conditions or the mode won't be as good as PSK-31.

In strong signal conditions, QPSK does outperform PSK-31 which has no error correction, but it is not better than PSK when the signal is weak or fluttery, and it was never very popular.

More complex modulation modes that use four or sixteen phase states or anything up to 32 frequency shifts (tones) are more difficult to demodulate accurately. As the number of possible symbols increases, the decision circuit that discriminates between the received phases or frequencies has a smaller window in which to correctly identify the transmitted symbol. When the received signals are weak, the received phase and amplitude noise randomly offset the phase and frequency of the received symbols making them harder to decode and ultimately causing errors when received symbols are detected in the wrong state.

The easiest way to show this is with a 'constellation' diagram that shows the phase states of the received QPSK signal **(Fig 2.2)**.

As the received signal becomes weaker the received symbols are affected by amplitude and phase noise causing a spread of the received symbols over time. Any symbol that moves so far out of position that it is detected incorrectly causes a 'bit error.' Fig (b) shows amplitude and phase noise caused by a weak received signal.

Fig (c) shows a similar amount of amplitude noise (bigger blob) and a significant amount of phase noise (rotation of the received symbols from their nominal positions).

As the symbol rate increases, the bandwidth of the transmitted signal increases. If multiple tones are sent concurrently this also increases the transmitted bandwidth. The wider receiver bandwidth required to receive the signal causes a lower signal to noise ratio, so faster more complex modulation schemes need more received signal, for error-free decoding, than simple narrow band modes. The signal to noise improvement resulting from a narrow band transmission is one of the reasons that weak CW signals in a 500 Hz bandwidth are easier to copy than SSB signals in a 2.4 to 3.0 kHz receiver bandwidth. The signal to noise improvement in decibels resulting from reducing the receiver bandwidth is the log of the bandwidth ratio, dB = $10\log(2400/500) = 6.8$ dB.

The other key factor is that most data modes, including CW, transmit at full power. An SSB transmission only averages 20-50% of the PEP (peak envelope power), depending on how much speech compression is enabled. So that accounts for another three or four dB improvement in the received signal to noise ratio.

Hellschreiber

The Hellschreiber was a mechanical "teleprinter like" machine developed in the late 1920s by Rudolf Hell, as a simple means of distributing text from central press offices to newspapers. It was also used extensively during WWII for non-secret military communications and possibly for sending secret message code groups encoded by Enigma machines. After that, it fell into disuse until it was 'rediscovered' by radio amateurs in

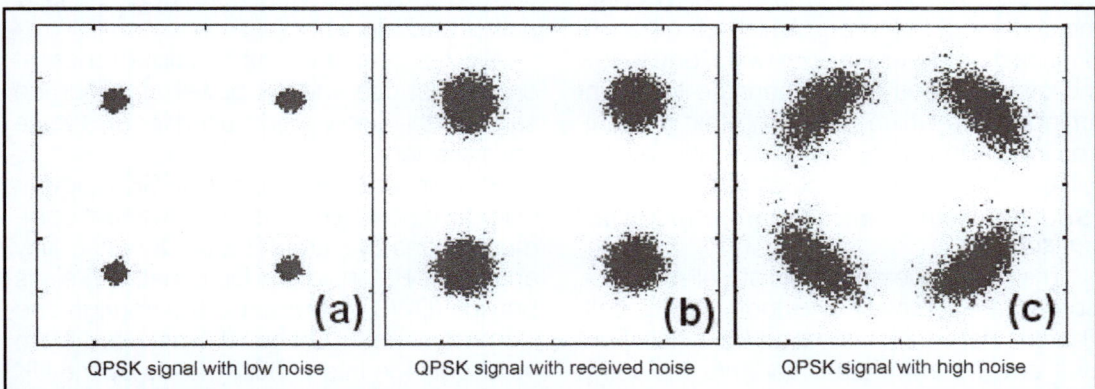

| QPSK signal with low noise | QPSK signal with received noise | QPSK signal with high noise |

Fig 2.2: QPSK with increasing received noise.

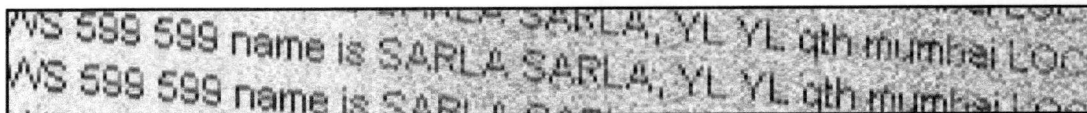

Fig 2.3: Received Hellschreiber text

the 1970s. While it is unlikely you will see a Hellschreiber transmission on the bands, it is an interesting mode from a historic point of view, and also because it is one of the two data modes which can be read without the need for a decoder. I have two 20m band Hellschreiber contacts from 2003 in my log, one confirmed.

Hellschreiber often shortened to 'Hell' is a facsimile mode, where characters are 'scanned' and transmitted. With the original mechanical machines, there was no need for encoding of the data and no timing or synchronisation. Characters printed onto paper tape were scanned with a light-sensitive 7-pixel matrix at a rate that creates 7x7 pixel characters as the tape passes through the reader. The pixels were transmitted using ASK (amplitude shift keying) at 122.5 bauds (about 25 words per minute). Those of us old enough to remember 7 pin dot matrix printers will be familiar with the idea. At the receiver end, a paper tape is fed at a constant speed over a roller. Located above the roller is a spinning cylinder with small bumps in a helical pattern on the surface.

The received signal is amplified and sent to a magnetic actuator that pulls the cylinder down onto the roller, hammering out a dot into the surface of the paper. A Hellschreiber will print each received column twice, one below the other. This is to compensate for slight timing errors which cause the text to slant. The idea is that if you can't read one line you will be able to read the other.

[Source: https://en.wikipedia.org/wiki/Hellschreiber]

There are several variants of Hellschreiber. Some were invented by Rudolf Hell himself. The original version is called Feld Hell. Feld Hell and its immediate variants including Slow Hell use ASK (on-off) keying. Feld Hell

80 (Rudolf Hell's last original Hell mode, designed in the 1970s) uses 2-FSK modulation, and the other FSK Hell modes use 2-DMSK (differential minimum shift keying).

If you look carefully at the waterfall display on an SDR or recent HF transceiver while it is receiving a Feld Hell signal, you can read the characters directly off the waterfall image.

MFSK Modes

MFSK (multiple frequency shift keying) is used for the many data modes that transmit multiple tones. Usually, the mode sends a single tone at a time. The information being sent determines which tone is sent next. For example, tone A followed by tone D may mean a data signal of 01100001. Tone A followed by tone E could represent a data signal of 01100010. Using multiple tones allows a high data rate per transmitted symbol. Each change of tone frequency relates to several bits. In some cases, a complete 8-bit data byte or an ASCII character is transmitted per symbol. Some modes transmit more than one tone at the same time, extending the possible combinations and increasing the symbol rate without adding more tone frequencies. Modes that only transmit one tone at a time can be amplified with non-linear class C power amplifiers.

Modes that transmit multiple tones at once must use a linear power amplifier the same as the 'linear amps' used for SSB voice transmissions.

Piccolo was the original MFSK mode. It was developed for British government communications by Harold Robin, Donald Bailey and Denis Ralphs of the Diplomatic Wireless Service (DWS), a branch of the Foreign and Commonwealth Office. It was used from 1962 until the late 1990s, mainly by the UK government for point-to-point diplomatic and military radio communications. The original version used 32 tones. Piccolo MK6 was a six tone Baudot coded system with a 180 Hz bandwidth. It was designed to be a faster replacement for RTTY. Piccolo MK12 was a twelve-tone ASCII system with a 300 Hz bandwidth. It could transmit at 110 bps.

MFSK8 and MFSK16 were developed by Murray Greenman ZL1BPU, for amateur radio communications on HF. They used 8 tones and 16 tones respectively. MFSK16 was the more popular mode. It can achieve a transmission speed of 42 wpm which is faster than most typists can type. The ability to keep up with your typing speed makes MFSK16 an ideal "chat" mode. It transmits the lowest tone then starts sending data, transmitting one of the 16 tones at a time. The receiver uses a Discrete Fourier Transform (DFT) to separate the received signal into frequency bins. The bin with the highest level is considered to be the received symbol. Forward error correction is employed because frequency drift can cause errors with this method. A phase detector corrects for frequency drift in the transmission. Murray Greenman also developed DominoF and DominoEX for NVIS radio communications on the upper MF and lower HF frequencies (1.8-7.3 MHz).

Olivia is another MFSK mode with forward error correction. It was designed by Pawel Jalocha, SP9VRC, to work effectively over difficult radio paths, especially those suffering from fading, interference, auroral distortion, flutter, etc. It can work when the signal is 10-14dB below the noise level, which means that worldwide communication is possible using low power. Olivia has 40 possible formats. There can be 2, 4, 8, 16, 32, 64, 128 and 256 tones, resulting in bandwidths of 125, 250, 500, 1000 and 2000kHz. This versatility is a drawback as it makes it difficult to decide which option to use. The most common formats are 500/16, 500/8, 1000/32, 250/8 and 1000/16. Data throughput is just under 60 words per minute at 1000/8. Olivia works by sending blocks of five 7-bit ASCII characters over two seconds.

MT63

MT63 is another MFSK mode. It was developed by Pawel Jalocha, SP9VRC, in the late 1990s, and is an excellent mode for sending text over propagation paths that suffer from fading and interference from other signals. It works by encoding text with a matrix of 64 tones over time and frequency. Although this is rather complicated, it provides error correction at the receiving end. In its most popular configuration, MT63 has a bandwidth of 1kHz and a throughput of about 100 words per minute. There are also 500Hz and 2000Hz variants, which are achieved by scaling the times and frequencies. On the air, MT63 sounds like a roaring or rushing noise.

The WSJT Modes

The WSJT (weak signal by K1JT) modes were created by Joe Taylor, K1JT an American astrophysicist and Nobel Prize laureate in Physics. Bill Somerville, G4WJS, Steve Franke, K9AN, and Nico Palermo, IV3NWV, have been major contributors to the development of WSJT-X since 2013, 2015, and 2016, respectively. WSJT-X has replaced the original WSJT program which was first released in 2001.

These modes have revolutionised amateur radio weak signal communications. Before the release of WSJT in 2001, EME communication was performed using CW and very occasionally SSB. The conversion to the far more sensitive JT65 data mode has allowed modest single Yagi stations to try EME. I managed to work, Russia on the 2m bands using a 160-watt amplifier and a single 10 element Yagi antenna. Not bad from New Zealand! Meteor and aircraft scatter contacts evolved from hearing 'Pings' and using very fast machine-sent Morse code, to being able to reliably carry out communication of callsigns and signal reports. Early morning meteor scatter contacts over the 3500 km path between New Zealand and Australia are common using MSK144 on the 2m and 6m bands.

WSJT-X, MAP65, and WSPR are open-source programs designed for weak-signal digital communication by amateur radio. Each of the programs includes several data modes tailored for different types of operation and band conditions. For example, WSJT-X includes 11 different 'protocols.' The most popular are FT8 which is used for weak signal contacts, primarily on the HF bands, but now also on the VHF and UHF bands. JT65 is used for EME (Earth-Moon-Earth) contacts on the VHF and UHF bands and JT4 is used for EME contacts on the microwave bands. Echo was designed to listen for the reflection of your signal from the Moon even if the returned signal is well below the audible threshold. MSK144 is a fast mode used to take advantage of brief signal enhancements from ionized meteor trails, aircraft scatter, and other types of scatter propagation. User messages are transmitted repeatedly at a high rate, up to 250 characters per second, to make good use of the shortest meteor-trail reflections or "pings." MSK144 replaced the old FSK441 mode that was available in the original WSJT program. It takes advantage of the improvement in personal computers since 2001 and has better error correction and sensitivity. It increased the transmission speed from 147 characters per second to 250 characters per second, which dramati-

cally improves the decodes from very short meteor reflections. The accurate decoding sensitivity improved from "a few dB above the noise," to around 8 dB below the noise level.

WSPR (weak signal propagation reporter) is designed to report propagation paths on the HF bands. It transmits a low power beacon and then listens for beacons from other stations.

Most WSJT modes use timed sequences of 15 or 60 seconds duration. WSPR uses a 2-minute sequence, but it typically transmits on only 20% of the sequences. i.e. it receives for five periods and then transmits for one period.

It is remarkable how sensitive WSPR is. The image shows the reception reports for a single 5-watt beacon transmission on the 15m band at 1:20 pm. I was impressed to see reception in the Canary Islands, via a polar path, in the afternoon.

WSPR messages normally carry the transmitting station's callsign, grid locator, and transmitter power in dBm, and with two-minute sequences, they can be decoded at signal-to-noise ratios as low as -31 dB in a 2500 Hz bandwidth. FST4W is designed for similar purposes, but especially for use on LF and MF bands. It includes optional sequence lengths as long as 30 minutes and reaches sensitivity thresholds as low as -45 dB. Users with internet access can automatically upload WSPR and FST4W reception reports to a central database called WSPRnet that provides a mapping facility, archival storage, and many other features. The map shows what the propagation is like across the world, providing an excellent guide to the bands that are open and the countries that are most likely to respond to your CQ calls.

[source https://www. physics.princeton.edu/ pulsar/K1JT/wsjtx-doc/ws-jtx-main-2.4.0.html#INTRO]

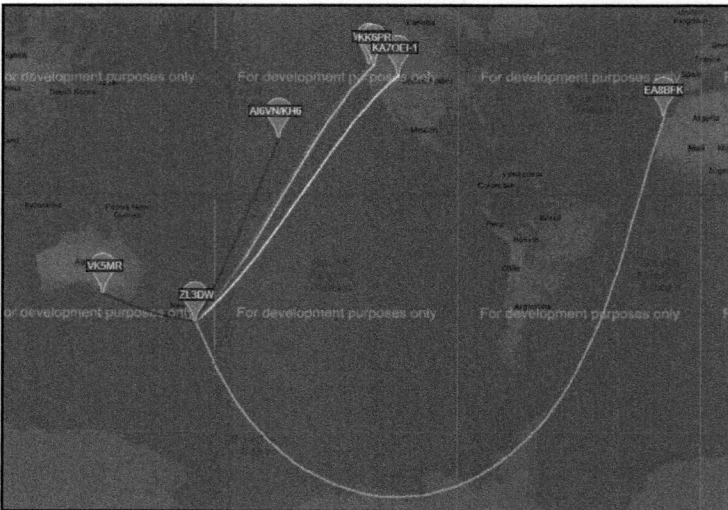

Fig 2.4: Reception report for a 5W WSPR beacon

The new Q65 mode is a new digital protocol designed for minimal two-way QSOs over especially difficult propagation paths. On paths with a Doppler spread of more than a few Hertz, the weak-signal performance of Q65 is the best among all WSJT-X modes.

Q65 uses message formats and sequencing identical to those used in FST4, FT4, FT8, and MSK144. Sub-modes are provided with a wide variety of tone spacings and T/R sequence lengths of 15, 30, 60, 120, and 300 seconds. A new, highly reliable list-decoding technique is used for messages that contain previously copied message fragments. Message averaging is provided for situations where single transmissions are too weak or signal enhancements too sparse for a signal to be decoded. Q65 will be especially good for EME where Doppler shift causes frequency drift and spreading of the received signal.

MAP65 is another EME mode. It works with hardware that converts RF to baseband, usually an SDR receiver. The combination of hardware receiver and computer software implements a wideband, highly optimized receiver for the JT65 protocol, with matching transmitting features using a standard SSB transceiver.

Serious EME operators use arrays of antennas to maximise the antenna gain and enhance the weak signals reflected from the Moon. They usually employ separate vertical and horizontal polarised arrays because Faraday rotation of the received signal as it passes through the atmosphere, coupled with the difference in received polarisation due to the relative position of the transmitting and receiving station on the Earth's surface, means that a signal that was transmitted using one polarisation (horizontal or vertical) might be received at any angle and may change during the contact. This cross-polarisation creates deep nulls in the received signal which can cause even a strong signal to drop out during the contact. Having both polarisations allows the receiving station to change to the antenna array that has the best signal. Or if the receiving station only has one

polarisation available, the transmitting station can change. Unlike satellite operators, the EME enthusiasts don't usually combine their antennas into a circular polarization array, because with the very weak signals reflected from the Moon, they can't afford the 3 dB signal loss that creates.

MAP65 can be used in both single-polarization and dual-polarisation systems but the real advantage is realised when two polarisation channels are available. MAP65 receives signals from both antennas and uses a phase delay to combine them 'additively' to improve the signal to noise ratio of the wanted signal. This capability provides a major advantage for efficient EME "Moonbounce" communication on the 144 and 432 MHz bands. The latest MAP65 v3.0 release includes the Q65 mode.

FT8

FT8 was added to WSJT-X version 1.8 in July 2017. FT4, a similar but faster protocol designed especially for radio contests, was introduced two years later in version 2.1.

As solar cycle 24 drew to a close and declining band conditions made long-distance contacts on the HF bands difficult, the new FT8 WSJT-X mode started to gain popularity. By the time the sunspots cycle reached a minimum, FT8 had completely replaced PSK-31 and RTTY. At many times of the day or night, FT8 was the only activity heard on the 20m, 15m, and 10m bands. The mode has an amazing capability to work when the bands are "closed." It also allows many stations with low RF power and modest antenna systems such as a small vertical or loop antenna on an apartment balcony, to make contacts all over the world.

The digital mode software (WSJT-X or similar) selects the normal operating (VFO) frequency for the FT8 mode, on the selected band. If the 'Split' operation option is selected, the software will automatically adjust the transceiver VFO while transmitting, to ensure that the FT8 audio tone being presented to the modulator is between 1500 to 2000 Hz.

This ensures that any audio harmonics cannot pass through the transmitter sideband filter to create false (duplicated) FT8 signals on the band. You should definitely select this option when you set up your WSJT-X for FT8 uses alternating 15 second transmit and receive periods. It transmits at 6.25 symbols per second for 12.64 seconds in each period, leaving 2.36 seconds for the decode at the receiving end, before the next transmit period starts. Each transmission sends 79 symbols (237 bits) of information. 77 bits containing the 3 'message type' and 74 bits of user data, plus CRC (cyclic redundancy check) bits, parity bits and Costas loop sync data. The bit coding analyses the message and sends tokens rather than ASCII characters. Common abbreviations, grid locators, and callsigns are sent in a compact format. For example, CQ is sent as a single byte, 02H. An indexing method compresses six-character callsigns, into four bytes of data and four character grid locators into two bytes of data.

The mode uses 8 tone FSK with a tone spacing of 6.25 Hz, resulting in a transmitted bandwidth of 50 Hz. The decoder finds and decodes all of the signals across the receiver passband, so it will display calls containing your callsign even if the FT8 "receiver" is not set to the tone frequency of the calling station.

An FT8 QSO consists of sending and receiving four or five pre-defined messages.

These automatically step to the next message as the previous message is acknowledged. Some software such as MSHV has automated this to the extent that the entire QSO can take place automatically. It will even queue up stations that have called you. The mode can transmit a 16 character message in each 15 second 'transmit period' so it takes two or three minutes to successfully complete a contact. There is a DX mode where a Dxpedition or contest station can carry out 'conversations' with several FT8 stations simultaneously. There is also a 'conversation' mode called JS8 using the JS8 Call software by KN4CRD, which uses the same signalling technique but does not require the preset messages. To some extent, FT8 is a victim of its own success. There is so much activity on the nominated 20m frequency that it is often hard to find a spot for your transmissions. The users need to spread up the band a bit, or a second VFO frequency could be added to WSJT-X. There is no rule that says that FT8 must be used on any particular frequency, provided it does not encroach on frequencies traditionally used for other modes. WSJT-X is not the only digital modes program to support FT8. You can also use MixW, Ham Radio Deluxe (Digital Master 780), or MSHV developed by LZ2HV.

Decoding Threshold

FT8 is a few dB less sensitive than the slower WSJT modes (JT65, Q65, and JT9) which have one minute transmit periods, but it can complete a QSO much faster. It has a decoding threshold of -20 dB, and this can improve to about -24 dB if the same data has been received before. It uses previous partial decodes to improve the decoding accuracy. A decoding threshold of -20 dB means that it can decode signals that are 20 dB below the received noise level. 12 wpm CW signals can be deciphered at about -12 dB. SSB speech at S2 on the signal strength meter, "barely readable, occasional words distinguishable," requires the signal to be about 3 dB above the noise level. WSPR signals can be decoded at 29 dB below the noise level. PSK-31 works down to about -7 dB and RTTY to about -5 dB.

These negative threshold numbers are a bit misleading. They are referenced to the noise level in a typical SSB receiver bandwidth, of 2500 Hz. However, the bandwidth of the transmitted signals is much narrower, and the digital mode software employs narrow filters or an FFT (fast Fourier transform), before decoding the individual signals.

Within those narrow filters, the signal to noise ratio of the individual data mode signals is much higher, and always above zero. For example, FT8 is stated as having a -20 dB decoding threshold in 2500 Hz. It has a transmitted bandwidth of 50 Hz so if we use a 50 Hz receiver bandwidth the 'process gain' resulting from the bandwidth ratio is 10log(2500/10) = 24 dB. The signal is 4 dB above the noise level within the decoding bandwidth. Using FFT, the signal to noise within the resulting output 'bins' are only a few Hertz wide, so the decoding signal to noise ratio is even higher. The very narrow bandwidth at the decoder allows the digital modes to work, even when you can't hear them on your SSB receiver or see them on a spectrum and waterfall display.

Finally

The data modes provide the opportunity to make worldwide amateur radio contacts in a different way. They are a relaxing way to operate, except while contesting! The signal to noise advantage of data modes over speech modes makes them ideal when received signals are too weak for SSB, or on paths suffering multi-path fading. FT8, PSK-31, and RTTY are great modes for hams that want to use the HF bands but can't erect large antennas or run high power linear amplifiers. You can work the world with QRP power and small loop, wire dipole, or vertical antennas.

'Digital mode' contacts qualify for most amateur radio awards including DXCC, WAS, WAZ and WPX. The UltimateAAC program is available from WWW.EPC-MC.EU. It combines the awards provided by several European data mode clubs and the 'FT8 Digital Mode Club' to provide hundreds of free, data mode, activity awards. The program reads your ADIF station log and automatically finds the awards that you qualify for. You click a link to apply and after an approval process, the awards are available for download as high-resolution pdf or jpg files. There are contests for many data modes, including worldwide contests for RTTY and PSK, and FT8 is now used on most DXpeditions.

Once the connection has been made between your computer and your transceiver you can easily try out a variety of data modes. As the bands improve over the next few years you will have a great opportunity to try something new. Find the right VFO frequency and make a CQ call on Hellschreiber, MFSK, or PSK-31. Or if you have a 6m or VHF DX station, point your antenna at the rising or setting moon and see if you can decode any JT65 or Q65 EME signals.

3
Logging Software

by Mike Ruttenberg G7TWC

In this chapter it is assumed you are either curious about, or already interested in, using software with your radio, and you're looking for either 'getting started' guidance or you're experienced already but want to know some sneaky tips. This chapter should show you the ropes and tips, so let's start from the basics for the uninitiated.

Much of this chapter will relate to contesting software since many operators have cumulatively many centuries of experience, so their learning and requirements are now standard in operating software, whether or not for contesting.

Whether you are looking to log your 'rag-chew' QSOs, a data mode enthusiast, love chasing 'DXCCs' (amateur radio 'countries'), or are a contest-winning operator, there is almost certainly something for you out there to cater for your needs. I will be using 'software', 'package', 'logger' and 'program', or a combination of these, interchangeably within this chapter.

There is no answer to the age-old questions, 'What software should I use?' or 'What is the best software?' It depends on what you want to use it for. You may also use a variety of packages for different purposes, e.g. contest vs casual logging or specific data modes.

I would also like to add that although this chapter mentions modes and much functionality, it equally applies to SWLs (shortwave listeners).

I want it all!
No single package does everything. We are blessed with a wide variety of niche interests in the hobby, but more than any one software package can keep up with.

The next factor is personal taste. Even for the same functionality, people are used to different 'look and feel' (colours, screen layout etc) or just familiar with a previous product and want more of the same.

Some packages allow you to rearrange the screen to personal tastes. In some packages, each component may be able to be moved around the screen to the operator's preference. Being able to move, enable or disable elements of the software can be very useful, e.g. if you only have a smaller screen.

Choosing your software
It all depends on what you are trying to do, so let's look at some basic questions to help you pick a best-fit solution to your needs.

Assuming you want to log your QSOs and you're moving away from paper logging, or picking a new possible logging program, basic questions need to be asked such as, but not limited to:

- Which mode(s) do you want to use when operating?
- Do you have a PC, a Mac or run Linux?
- Do you want to (and can you) connect your rig to the PC for it to control the rig e.g. frequency and mode logged automatically?
- Do you operate on data modes? And if so, which ones?
- Do you contest frequently, or never?
- Do you have disability/accessibility requirements?

As no package does them all, it's a best-fit solution you need to look for, and you can always use different packages for different purposes.

Cyberphobia

For some, the time-honoured paper log is their preferred method but see the list below for the advantages of using a computer instead. You don't need specialist knowledge for any of the software packages and there is a lot of assistance for 'newbies'. Often setup is a one-time operation and thereafter, generally, works using the configuration you set (or that comes as the default) unless you change it.

Advantages of using a computer in Amateur Radio

Below is a list of a small number of the many advantages of using a computer in amateur radio, in no particular order. Please note that not everything applies to, or is present in, every package.

- Many data modes use the computer's soundcard to do all the hard work for you, which previously required external interfaces. e.g. for data modes,
- Less chance of mislogging calls due to callsign database making suggestions.
- Live feed of 'DX cluster' data of stations on the air and their frequencies, fed from the internet.
- Visual 'bandmaps', telling you who is on what frequency, fed from the DX cluster.
- Visual alerts telling you if a station is a new country, state etc.
- Real time 'dupe' checking (alerts you whether you have worked the station before).
- Computer-generated and decoding of many data modes, e.g. SSTV (Slow Scan TV), RTTY and many others.
- Displaying the 'greyline' (visual indication of where the sun rising/setting) on a real-time map. 'Greyline propagation' is useful to contact places along that line, as propagation is enhanced between areas along the 'greyline'.
- Silent operation: Operating can be purely visual, using data modes, pre-recorded voice messages or code readers, which allows operating in silence.
- Computerised assistance with decoding CW ('carrier wave', AKA Morse Code).
- You can export logs for submission to logging sites and/or awards sites.
- You can export and/or print QSL card details from some packages and much more besides.
- In addition to the above, contest functionality can include, but is not limited to:
- Networking PCs together for multi-operator contest or event stations .
- Cluster spots inform you of multipliers/ new countries to work, the frequency they are on and (for CW) how strong they are where they were heard.
- Support for pre-programmed 'macros' (predefined messages, including audio, attached to a key press)
- Auto-repeat of CQ calls, which repeat after a customisable period of time, saving time and energy.
- Validation of the contact against the rules, e.g. in contests where you can only work another country/prefix/zone etc
- Export of logs to standard formats ready for upload to online logging systems, contest adjudicators and awards systems, QSL card printing software and awards programmes.
- Ability to set the operator callsign. This enables a single computer with different settings and macros to be shared between different operators.
- 'Callsign stacking': Where another operator listens to your frequency and enters who else is calling you, for faster logging in a pileup.

This is not a finite list - there is much more.

I hate contesting!

General 'DX' modules for general logging are almost invariably included, and the other features come in useful too, such as checking previous QSOs, ability to print cards, submit logs for awards, prompting you with operator names, telling you distances and bearings, decoding data modes and much more.

The software packages that support contesting are generally fantastic general logging programs, and many support additional modes.

Computing Power

Do I need a whizz-bang supercomputer to use software packages? No you do not. Some software uses modest PCs for our modern day and age. Some software runs fine on a PC from the mid 2000s, so long as it runs a form of modern-ish operating system. On PC, generally Windows 10 is recommended (at the time of writing).

For most purposes a modest PC or laptop will do, though the faster and more modern PCs do the same thing faster and allow greater sophistication. You may also find that the ports (e.g. serial, LPT ports and FireWire) are no longer supported or present, though adaptors and interfaces for your rig or PC may overcome this.

If you're going to use a (much) older PC I would suggest it is a dedicated machine with nothing other than the operating system installed on it and the logger. That way there is less strain on the system while trying to run the operating system and the logger (which tries to harness more computing power that the PC might not be able to provide). Software writers often try to keep up with technology, though some of the technology we use doesn't always keep up with the demands of the software.

Tip: For Windows users: A common issue encountered is that the software package needs to run and/or be installed with Administrator privileges, so check when installing that you install as Administrator. Usually right click on the installer -> select Run As Administrator. This should do the trick in many cases.

What about Macs?

Macs are not well supported by developers due to the prevalence of Windows-based PCs. The advice and tips below though apply equally to Mac packages as Windows ones.

Users have been known to run packages successfully under virtual machine instances of Windows.

One package for Mac is SkookumLogger (https://www.k1gq.com/SkookumLogger/). Another is MacDXLogger (https://www.dogparksoftware.com/MacLoggerDX.html). WSJTX for has a Mac version, for some of the newer data modes. Other packages may be available.

Linux

There are logging and contesting tools for Linux too. However, it's a specialist operating system that, although loved by many, isn't a well-supported platform for radio users, and many radio packages are supported under Wine. The advice and tips in this chapter apply equally to Linux packages as other operating systems.

Logging Software

Does software ensure accurate logs?

No, you still can mistype a call or exchange, or even if you get all the details correctly there is no vouching for the other station copying and/or logging all of your details correctly.

Which is the best software for logging?

How long is a piece of string (or antenna wire)? This is the holy grail question of logging, and there is no unique answer. It depends on your needs, with a rough and ready set of questions as follows:

- What mode are you operating with? Some packages support some modes better than others, e.g. CW/SSB vs data modes, and are using specialist data modes, e.g. for meteor-scatter, moon-bounce or monitoring beacons?
- What operating system are you using? PCs are overwhelming better supported.
- What contest(s), if any, are you operating in?
- Are you a casual operator looking to add to your DXCC count with no intention of submitting a log so don't need scoring or serial number logging?
- What country do you live in? The software is most often in English, but some software exists in French, for example.
- Cost: Do you mind paying for the software? Some is free, some comes with a cost.
- Do you have internet?
- Do you want to include cluster spots from the internet, or not at all?
- Do you want to read and send the CW manually or via the computer?

As you can see, this is a significant list of variables, in addition to the previous question lists in this chapter, and it is sure to grow with the advent of future technologies.

How does the software stay up to date?

Some software automatically prompts you to download updates when you start it up. Others require you may to manually check for updates. If you subscribe to the reflector/user group for the software package (e.g. Win-Test), you will see notifications that there are newer versions available to download.

If you pay for your software, you may get a period-limited right to updates e.g. updates for 1 year, or for life. Check your terms of the software licence.

Note: You may have to pay for major version updates on some paid software packages. Some packages are free, become paid products, and may even return to a free model.

How do i get new prefixes, countries and/or multipliers in to my software?

When countries split apart, new countries are created/declared or countries issue new prefixes, the software doesn't know about this, so it refers to a file telling it which prefixes (or specific callsigns) belong to which country.

Over the years the software developers have settled on a handful of formats of the 'country file', which can be quickly and easily plugged in to the software package so that countries are recognised when a callsign is entered. This also helps the scoring in contests if the prefixes are important, as they need to be recognised to be scored correctly.

Although most packages ship with a basic country file included, you may need to download the most recent country file, known as 'CTY.DAT', though the actual filename or format may differ depending on your logging software package. You can get a copy from http://www.country-files.com. Some logger software packages include built-in menu options to grab the right file from the right location, update it and put it to use. Check the instruction manual for your preferred package how to do this, or if it is supported. If not, doing it manually is usually supported.

Many contesting and software email reflectors publish that a new country file is available for download.

Does the software support multipliers if i am in a contest?

Many contests and awards have different multipliers ('mults'). Sometimes it's the prefix of the callsign, sometimes it's the DXCC entity (country or region/province etc within a country), CQ zone, ITU zone, island ('IOTA') reference, province, state, region, county, postcode prefix, specific club membership number or whatever. Every contest or award has its own rules and what, if anything, counts as a multiplier.

In addition to countries, some contests want to split up a country into different multipliers, e.g. county, state, oblast, canton etc. To this end, they may use configuration files that splits up the country or region into the required mults using a .DOM and/or .SEC file. These translate callsigns into the relevant category for the software to recognise the relevant states/provinces/ call areas/counties etc for correct scoring. You shouldn't normally need to edit these. Different software packages treat these differently, and sometimes not at all.

Check in the contest rules and/or logger's instruction manual whether a particular package support a particular contest. If the contest is supported, it probably handles the mults for you, and usually the mult files (.DOM or .SEC files) are bundled with the installation package. Many packages support a wide range of contests, listed in their help files.

Consult the manual for your preferred package about how to invoke the relevant mults files in your chosen package. Usually selecting the correct contest is enough. Occasionally you may have to tweak additional configuration settings.

In the rare instance that you need additional or updated .DOM or .SEC files, these can often be found on the internet or by asking other users on your chosen software's support email mailing list forum. Someone usually had the same question and can provide you some helpful guidance.

On-screen indication of mults still needed

Some software lets you view which countries or other mults you have previously worked and which ones are still needed, and some software will tell you on screen that the station you are working is needed on other bands, e.g. you are operating in a contest and have worked GJ6YB on 10m SSB but need GJ (Jersey) as a multiplier on 10m CW and 15m SSB. Each logging software package displays worked mults and needed mults differently, where supported. This is a trade-off between the best use of computer screen space, software power, software development effort, software cost (if any) and, as always, operator skill to invoke the relevant window and monitor it so that each multiplier gets worked in good time, for maximum points.

This is not included in the grid at the end of this chapter, as it depends on the contest whether mults are supported. In some packages, the on-screen mults window can be hidden/moved or made more prominent, based on the user's preference.

In any case the adjudicators for the contest will rescore your log, whether or not your multipliers are recognised and scored correctly. However, the best way to maximise your score is to work as many multipliers as possible, so being alerted to what mults are available to work (whether via the bandmap colouring and/or the mults window), is the best way to incentivise you to work them before propagation changes or they change frequency before you get to work them.

Tip: Knowing which bands you need a mult on is useful as you can sometimes request a station to move from one band to another by simply asking them to change bands ('QSY') to a band where you still need them as a multiplier. They may have a station on that other band already, or they may be prepared to change bands for you and meet you on an agreed frequency. The point is that seeing what mults you still need from a visual prompt, you can raise your score (or country tally or whatever) which you wouldn't otherwise have known you needed.

The mult i worked isn't recognised/ is incorrect. What can i do?

The most common issue is to have a callsign that isn't recognised as the correct DXCC entity. e.g. for TO callsign prefixes as all French overseas territories' special prefixes have the same prefix, 'TO', instead of FM, FG, FJ etc. the simplest solution may be to log the call at the time of working it, or if not accepted, force log it (see your manual how to do this). Some stations aren't where their callsigns historically suggest, e.g. KH6M is in Florida (historically W4 prefix) instead of Hawaii (KH6 prefix).

The most common reason for not recognising the country correctly is that you didn't have the most recent country list (CTY.DAT). You can update the country file and rescore the log.

Sometimes the country file still won't know what country or state the station is for, and that's okay. A later update to the country file may resolve this, just not right away. Even if you are using outdated country files, the adjudicator for the contest (if you send in an entry log) ignores claimed scores and re scores the contest, so you do not have to be (and rarely are) 100% accurate at the time of submission. Your log will be recorded anyway, as long as you submit your log to the contest organisers. Check the rules for the contest how to submit your log.

Yes, it is frustrating that a callsign isn't recognised correctly, or a callsign seen on the Cluster shows a great new country, but, sadly, this is often a misspot, i.e. the spotted callsign wasn't correct e.g. E51AA (Cook Islands) instead of ES1AA (Estonia).

What is, and do I need, a partial Callsign database file?

If you type in part of a callsign into the software, and you missed some characters in the full call, you can get help with identifying who the station is using a look-up file of known callsigns.

Basically, it's a list of known callsigns worked my major operators (usually in contests) and put into a list that is recognised by many software packages. You don't have to use it if you don't want to. For reference, it's commonly known as 'master.dta' or 'SCP' (Super Check Partial).

Here's how it works: Say you type in 'G7TW' into the callsign field. The software looks for these characters and shows G7TWC and DG7TW in a dedicated window on your screen. You can either pick one of the suggestions and correct the callsign in the callsign field, or you can ignore the suggestions

completely. It isn't a replacement for using your ears, but it does helps to guide you to a full callsign or to jog your memory, using previously-known callsigns that part-match the one you entered.

Another partial match function that some packages support is 'N+1'. For example, you type in G4ZVW. This suggests G3ZVW or G4ZVB. Unlike the previous suggestion functionality, it isn't filling in a missing letter - it is taking the callsign you entered and seeing whether, by changing 1 letter at a time, it might be a match to a known callsign, and it suggests alternative options for you to make a correction, or you can ignore the suggestions and log what you heard.

If your software supports partial callsign databases (see your package's documentation), then, if supported, it will download it and install it from http://www.supercheckpartial.com/. Some packages may require you to do this manually. SCP is usually updated about four times a year, and often includes the logs from major contests such as CQWW and CQWPX. Your package may have the SCP/callsign history update functionality already built in.

Note: You cannot view the SCP file with a text editor as it is encoded.

Tip: You can edit an SCP call file or build one yourself from your previous logs. Some packages can scrape your old logs so you don't need to do this yourself. However, if you want to construct a file yourself, MEdit software is useful to do this. (available from http://www.dxatlas.com/MEdit/). It generates a '.dtb' file, so check whether your package supports this. Trial and error (and patience) may be required.

Tracking band and frequency ('cat control')

Depending on the model or radio and which software you are running, you can interface the radio to the software so that, using data sent from the rig to the PC, it knows what frequency (and even what mode) you're on. This then goes in to the log entries, so that the contacts are logged against the correct band and mode.

This is usually done via Computer Aided Transceiver ('CAT') control. Basically it's a serial port (on older radios), USB port or alternative cabling system that interfaces the rig with PC. It is usually on the back of the rig and plugs into the PC. For older rigs you may require a module or specialist 'CI-V' module or specialist cable to do this.

Once connected to the PC, select the rig from the logging software (many packages support a wide range of rigs) and then the relevant information is sent to the rig and back by the software, which has been written to interface with the rig on many pieces of functionality e.g. PTT on/off, frequency, sideband/mode, RIT shift, rotator info and quite a bit more. What functionality is supported depends on the software and the capabilities of the rig. Check the software package, and

check your rig, as not all functionality is the same on all rigs. The settings will be rig-specific, and all packages expect the same settings when using the same rig, so they know how to talk to the radio.

The most common errors are the settings used. Check on forums and other users which settings are preferred for you rig.

Also check if you need to configure additional settings to enable CW, as this often goes alongside CAT control, so the rig knows when and how to send CW from the computer.

CAT control is vital for clickable bandmaps populated by cluster spots. When you click on a spot in the bandmap it tells the rig to change mode (if needed), select the correct band (if needed), go to the frequency, and then put the callsign on screen, all based on the clicked spot.

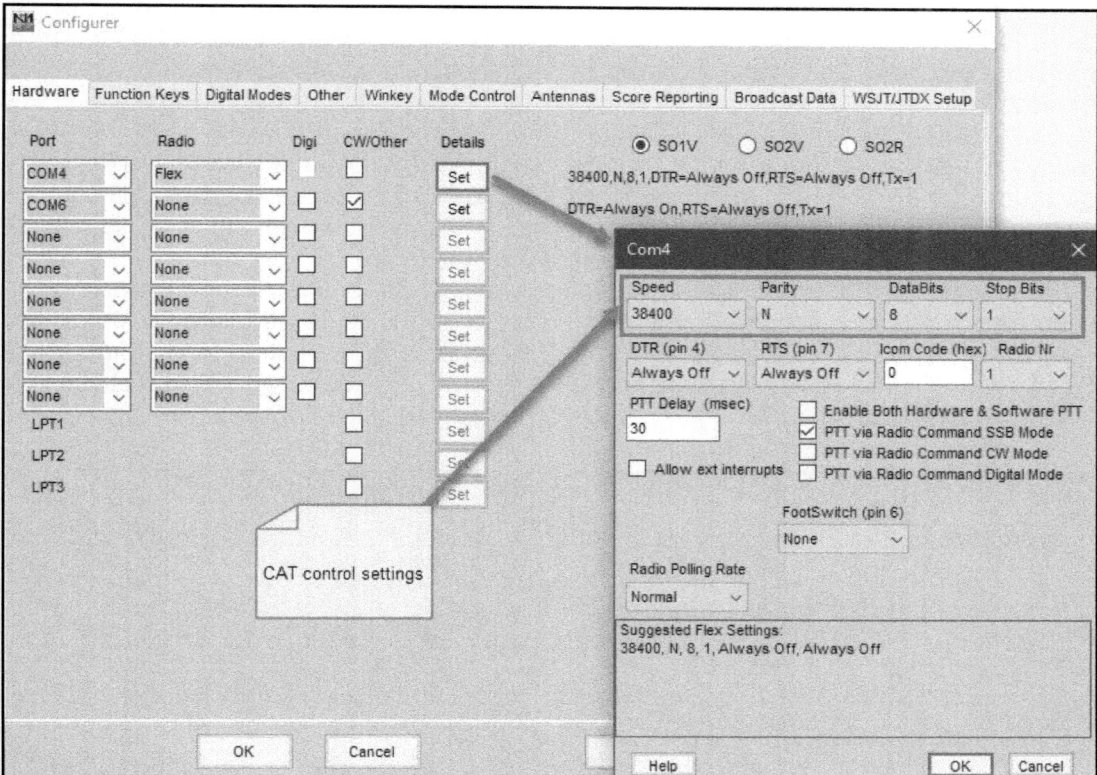

Fig 3.1: CAT control configuration settings

General Software Tips

What are Macros and why are they needed?

Macros are basically message shortcuts assigned to a key or combination of keys. They are useful for data modes, CW and SSB to save time and keystrokes when sending complex messages. Macros are powerful tools that can be used to send a message and a custom item e.g. your name/callsign/ serial number/locator or whatever.

On any program that supports macros, the macros can be personalised however you prefer, assuming the software supports macros.

There is a convention that most packages share a general set of common keys e.g. F1 = CQ. This is good news for most of us, but beware that not all macros and keypresses are the same in all packages. e.g. CW keying speed up/down.

It is not suggested that you change the basic key assignment, which is commonly (but not always) as follows (on SSB/CW/data modes such as RTTY):

Every software package has a different way of setting up the commands in their own way for the software to recognise it, so read the instructions and don't be afraid to ask for help.

It is suggested that you tackle the modes separately, to avoid confusion as the commands may be different for, say, CW and RTTY, especially if you are using data modes through MMVari or MMTTY. Once you get one

mode configured as you like it, then tackle the next one and so on.

For SSB, if using pre-recorded sound files, you will need the sound files already recorded to be able to chain them up into usable sequences. More on that later.

The next step of this is to bundle macros together in to the powerhouse that they can be, using 'ESM'.

What is 'ESM' and why you might want it

ESM stands for 'Enter Sends Message'. It is mainly for use in contesting. Think of it as a shortcut for many key presses.

Being a shortcut, ESM means you don't have to find and press the key assigned to a macro when you're calling CQ. It's much more useful for 'run' (CQing) stations, but still useful for search-and-pounce (S&P) stations.

ESM visual prompting

Software packages normally highlight which buttons are those that would be sent next during a contact when using ESM.

In the example shown in **Fig 3.2**, G3ZVW has replied to my CQ, I have typed his callsign. As highlighted, ESM will next send the contents of F5 (acknowledge G3ZVW) and then F2 (my exchange) the next time I press the ENTER key.

With ESM on, you are not expected to press the F-keys, although they are still available for you to use as normal. You could press F5 or interrupt your ESM flow to send a custom message from another F-key, or you

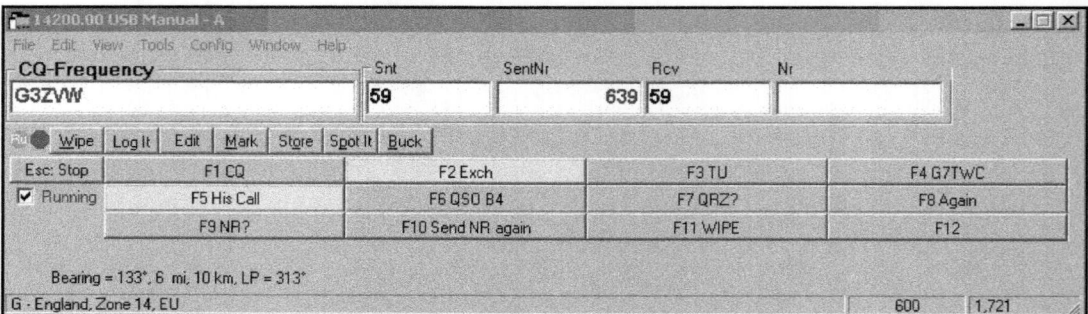

Fig 3.2: Stages of communication.

could use your Morse key at any point. Using ESM doesn't stop you doing what you want, but it does simplify the basic sequence of tasks and prompt you what is coming next.

ESM also works in SSB (see below about macros and voice files), assuming you have a sound card or DVK.

ESM and Macros for CW

What goes in to your macros is a question of taste, but for the uninitiated you may want a template to work off.

Whilst every package has a different 'dialect' of macro, the idea is generally the same. Consult the software manual for the exact commands for your package.

The key thing to do is to try it out. Note that N1MM+ doesn't support side-tone keying without being plugged into a rig (a frequently asked question) so I thought I'd mention it.

What DVK Is And Why You Might Want It? 'DVK' is short for Digital Voice Keyer. Basically it's a module or specialised hardware that you record or upload your voice into, and at a key press either on the module or via the computer it plays back the recorded audio file through the rig. Consult your instructions on how to upload or record the audio and set up the playback.

DVK does not use the computer's soundcard to play back the file. It's basically a soundcard in a stand-alone box and it does what it's meant to very well. Various manufacturers make these for ham purposes.

Voice Macros for SSB

Why use macros for SSB? To reduce the workload on your voice. It is highly recommended rather than straining your voice over many hours in a long contest. It is also useful for operators with speech disabilities. It can also help those that don't speak much English because they can use pre-recorded messages.

To do this, some operators use a soundcard to send the message for SSB contacts instead of using DVK. Given this functionality is built into all modern PCs, it is cheaper than

buying and setting up a dedicated DVK.

In order to send audio files from your keyboard using macros, the idea is that you need three things:
1) Prepare (record) the sound files you will need.
2) Have a lead from the PC's headphone/Line Out socket/USB to headphone converter (if you have no headphone socket), which you wire into the rig's PATCH port (if it has one) or mic socket.
3) Set up the macros in the software so it knows where to find the audio files.

Once configured, when you press the relevant key. The macro makes the radio go to transmit, any audio from the PC will be played and sent out on the air, hopefully with an intelligible relevant message. Note that this will also send any PC alerts or system start up sounds etc, so turn those off wherever possible. Instructions how to disable this are below.

Note: To pipe the audio from the PC to the rig, some rigs work with audio cables, some with USB, others with FireWire/Thunderbolt or Ethernet cables (or whatever technology comes along to replace it). The idea is the same. You're piping audio signals from the PC to the rig. How each rig receives the audio differs from rig to rig. Check your instruction manual and/or research online or forums for assistance.

While the setup advice for the audio files is in the next section, you still need to configure the software's macros to call the relevant .WAV-format sound files. For that, consult your logger's manual.

Playing sound files from key presses works whether you use ESM or not, so works for casual operators too.

TIP: On Windows 10, right click on the speaker icon in the bottom right hand corner on the Taskbar, select Sounds, ensure you select the Sounds tab, change Sound Scheme to 'No Sounds', press Apply, press OK. Now Windows sounds are all muted. If you require alerts for emails etc, don't change this setting but then when you are operating, you risk your 'pings', 'dings' and 'bloops' being played on the air too.

Recording and setting up voice files using your computer's soundcard

If you're thinking of using your computer to play your SSB macros, you will need a sound file for each part of the macro you are trying to send. You need to have recorded your sound files in advance.

You can usually get away with only a small number of files such as for CQ and sending your callsign, and then configuring the macros to play them. For example, you will need a CQ file (saying something like 'CQ Golf Seven Tango Whiskey Charlie') and a MYCALL file (saying something like 'Golf Seven Tango Whiskey Charlie'). The file normally needs to be an uncompressed Wave (.WAV) file format, and given a name, e.g. CQ.wav

Audacity (free from http://audacity. sourceforge.net/) or other sound recording and editing package is adequate. Remove all leading and trailing spaces from the sound file so there are no unnecessary silences, and 'normalize' the volume.

Next, you need to assign the macros for the relevant keys to the sound files, e.g. F1 to play a CQ file. Consult the manual for your package on how to set these up. e.g. In N1MM+ using Windows 10, the path for your WAV files is usually C:\Users\<Windows username>\Documents\N1MM Logger+\Wav\<yourcallsign>\

Next, test that the playback works as expected without connecting the PC audio to the radio yet. Do this by pressing the macro key, e.g. F1. You should hear the relevant recorded sound file play from the PC speaker. If you have CAT control enabled, the radio will go into transmit, though at this stage there won't be any audio. This is expected behaviour. We're just testing the macros successfully play the correct file from the PC at this stage.

The most common errors why the file doesn't play are because the file isn't in the right location or the filename that the software is looking for doesn't match the filename in the folder. Also check the volume setting on your computer is at a comfortable volume and the PC's soundcard/ speaker isn't muted.

When you hear your sound file play successfully when you press the macro key, next, plug in your cable from the headphones/line out socket to the Patch port, USB, mic socket or whatever technology you require to pipe the audio from the PC to the rig. Turn down the power output on the radio to zero and/or plug a dummy load into your rig. Next, press F1 to send your recorded CQ message. If the Monitor function is enabled and set at a comfortable level you should hear your recorded CQ message playing and your radio has gone into transmit mode. If it doesn't then check your CAT control settings since the playback function without the radio was tested and working without the radio connected. Also check the patch cable/USB etc cable that you used is plugged into the correct socket and doesn't have any breaks in it.

Common Sound Errors

If you receive reports of distorted audio on playing a sound file from the PC, you almost certainly may need to adjust the play volume on the PC. It's very easy to overdrive a transceiver's microphone input from a perfectly clear-sounding audio file. The rig is sensitive and will pick up what it thinks is loud audio (and which we think is perfectly acceptable).

Sometimes the sound file is distorted or too loud when it's recorded. Listen to it in a sound player (not through the monitor facility of the rig) to determine if it is distorted. If it is, re-record it. You may be able to adjust (or 'normalize') the volume within the recording software.

Voicing Callsigns with letters and numbers

Some software packages can send audio letters and numbers for you, e.g. the station you are working's callsign and your sent exchange. Although this is tricky at first to set up, this is a decent option. It usually sounds like robotic because the intonation is often wrong, but it works.

The advantage of using this is that it gives options if you have limited vocal capacity, a sore throat or want to operate silently (since the PC plays the files, and a microphone becomes obsolete). The disadvantages are that it can fiddly to set up, often sounds unnatural and doesn't cover impromptu conversations or non-standard replies.

TIP: N1MM+ has a section dedicated to this in its manual labelled 'Voicing callsigns…' at https://n1mmwp.hamdocs.com/setup/function-keys/#voicing-call-signs-serial-numbers-and-frequencies.

TIP: An excellent guide to recording natural-sounding letters and numbers, and how to optimise the sound files in Audacity is at https://n1mmwp.hamdocs.com/setup/function-keys/#recording-letters-and-numbers.

Auto-sending an operator's name

There are some contests where you need to send the operator's name (or you may just feel like sending it). Automated retrieval of the name can be done with a 'friends' file, usually named 'friend.ini'. Each software package, if supported, will need to be set up separately, so please consult your manual.

A friend.ini file that is generally available and quite comprehensive can be found at http://www.af4z.com/ham/.

In effect, you can create or add your own friends file by taking the general one and copying the layout and editing the contents using a text editor.

An example would be a combination of macros that when strung together make up 'G3ZVW de G7TWC GA STEVE UR RST 599 HW?' where 'STEVE' is inserted because he was inserted from a {NAME} macro.

The {NAME} macro matched the callsign G3ZVW, which returned the name, and so this was output.

Sending the time

Some contests (e.g. the BARTG RTTY) require you to send the time in UTC as part of the exchange. This is usually a macro, and may differ from package to package, so please consult your manual for the macros required.

Rotator Control

Some packages can control antenna rotators. You need to investigate this based on your requirements and the model of rotator you have.

SteppIR antenna control

Some packages can communicate with a SteppIR antenna controller. You need to investigate this based on your requirements. A good starting point is whether the software supports CAT control. If yes, then can it send this info to the controller unit? It also depends on the model of rig you have.

Operator Statistics

For multi-operator stations, some software packages support operator statistics. If you can set the operator's callsign in the software (separately from the callsign being used on the air), then you can create statistics based on an operator's on-time, to see the rate and how some operators performed or how the band shaped-up during their turn at the helm.

Naturally enough, to do this the software needs to know which operator is on the air. Check your instructions of your package, if supported, on how to set up, and specify the current operator's callsign.

Personal Audio Files

Some software packages can be configured to use specific audio files, depending on which operator is on the microphone. As long as you tell the software which operator is on the air, the files that correspond to the operator are played.

For example, when G7TWC is operating, pressing F1 may play G7TWC's audio file for CQ. When G3ZVW is operating, it would play their recorded CQ file, even though they also press F1. In short, setting the operator changes the context of various keys so they play the relevant sound files.

Some software allows on-the-fly recording of audio files, that is to say, an operator can modify and rerecord their sound files while operating. Check your package's manual whether this is supported.

Some radios allow recording and playback of recorded audio files too. Check your rig's manual to see if this is supported and how to set this up. This is independent of the software package you are using.

Data Modes

There are various software packages for data mode operation. Many modes exist, new ones are created from time to time and there are variations on themes (e.g. FT8, FT4, JT65, JT9, BPSK31, 63 and 135, various MFSK flavours and others).

In short, you can use your PC to send and receive various datamodes. Sometimes logging packages do this, other times you require additional software to assist with this task. See the dedicated chapter in this book related specifically related to data modes.

DX Cluster (Formerly 'Packet Radio')

Operators and monitoring systems gather data and allow others to be informed what callsign, modes and frequencies other operators are on. These pieces of information are called 'spots' (formerly 'packet spots', which required dedicated hardware).

The over-the-air Packet Radio system has dwindled and been ostensibly superseded by internet-based nodes. This brings you the cluster data effectively in real time and in greater volumes than ever before. It can populate a 'bandmap' (a panel on-screen showing you what stations are on what fre-

quency in frequency order), and can often colour-code the spots of new countries to work, zones etc.

To connect to a cluster, if your software package supports it, you connect to a cluster 'node'. This feeds the software with the data posted by other stations around the world. You can customise which modes you want to receive spots for, zones, where the station has been heard (e.g. Europe) or other personal preference.

On CW, the Reverse Beacon Network ('RBN') is a system of monitoring stations that decode CW and post what is heard into a cluster network, which some cluster nodes may relay (note that not all do as standard). Using/disabling this functionality can often be configured using command settings in your cluster window within the logger (or externally if you are using an external cluster program).

Once connected to a cluster node, if you want to, posting spots of who you hear/work can now be done en masse from your package merely by enabling a setting that posts all stations that you work in 'Search & Pounce' mode. That is to say, any QSOs you made where you weren't the station calling CQ.

Why would you post spots? If you work someone, it is a courtesy to others to inform them who is on what frequency and mode, so others can work them too.

Many logging packages include DX cluster functionality. Consult your manual for cluster access, to see if accessing the DX cluster is supported.

Advantages:

- You can find extra stations/countries/ zones or multipliers to work, boosting your points if you are in a contest
- You can find extra stations to work, e.g. if the band seems dead, you may want to listen on a spotted station's frequency and see if they appear out of the noise for you to work them when you wouldn't otherwise have known they were there.
- Some software packages populate a bandmap (see below) and may allow

you to click on the spot to take you to their frequency and pre-fill the callsign box ready for you to work them

- Spots can be posted in near-real time for you to see

- You can ask your software to retrieve a backlog of spots quickly, e.g. get the last 200 spots from the network, or set filters on the spots you receive. e/g/ on tell me spots on my chosen band or mode, or country/states of interest.

- Even if you can't hear a spotted station it can indicate that propagation is around, e.g. on 6m if Belgians are hearing Italians it may be possible that the Sporadic-E cloud will reach UK shores too.

- You can set filters on the spots so you only receive the mode/band/multipliers you want e.g. you only want 80m stations from G working CW. It's fiddly to do but it's possible. (see the Bandmaps section below.)

Disadvantages:

- Spots can be posted from anywhere, e.g. a US stations spot a VK (Australia) on 80m but you're in England on working on 80m in the daytime. You have no chance of hearing the VK so it is not helpful to see this type of spot.

- Being connected to the cluster is no guarantee of you being able to hear the DX.

- In remote locations, you may have no access to the Internet, even using your phone as a hotspot, and/or the cost may be prohibitive.

- The language used by the DX Cluster for you to set filters, send or force an update of spots can be fiddly (see the Bandmaps section below.)

- DX Spots can be wrong - callsigns incorrectly logged (sadly all too common), zeros and Os transposed, letters typed in the incorrect order, etc. Watch out especially for missing '/P' in spots

which can affect your score as portable stations are worth extra points in some contests. In short, rely on your ears, not on the screen.

Using the DX Cluster over the mobile phone network

Now that many mobile phones have high speed data functionality, the internet is available wherever you go (subject to coverage), so you could use your phone as a wifi hotspot for your computer to link to using 'tethering', and then display that data within your chosen package. For more info on setup of tethering, consult your phone manual or search on the internet.

Note: There may be costs associated with data usage, so consult your provider, know your tariff and be especially careful when abroad as data charges can be high and may not covered by your tariff in the location you are operating from. Check your 'roaming' policy for data charges and usage limits when travelling.

Filtering The DX Cluster spots

Setting filters in a language that the cluster understands is difficult for the unfamiliar user Powerful software to do this is AR User software (AKA 'VE7CC'), available from www.ve7cc.net. Once installed, you connect AR User to your chosen cluster, but now you can configure the bands and modes you want and don't want, or which continents or countries you want/don't want to see spots from. This is done via a graphical interface, so it is much easier to use. Once configured, many packages work well with the data output from AR User.

Bandmaps

If you are connected to a DX cluster you receive the spots data. This gets used to populate a graphical representation of the band you are on, known as a 'bandmap'. The spots are clickable, so when you click on the spot on the bandmap, the rig changes to the required frequency, generally populates the log window with the callsign of the station you clicked on, and you may get to work them. If you are operating in a contest, the bandmap is generally colour-coordinated to indicate which stations are multipliers, so you can choose to prioritise attempting to work those stations first.

CW Readers

CW Skimmer

'Skimmer' software allows you to use the soundcard of a computer to decode the entire audio passband of the rig or SDR unit into intelligible information, i.e. callsigns and/ or exchange info.

Basically, you connect the audio from the rig or SDR and it decodes the CW for you. How successfully it does this depends on the quality of audio, the rig, and the passband. If your passband is 100Hz, you may only get 2 or 3 stations decoded, but if you pass 100Khz of 20m in from a Software Defined Radio (SDR) you may get hundreds of stations decoded.

The skimmer is only as good as the audio passed to it, and you being able to 'view' the signal you want to contact (amongst many others also being decoded simultaneously). The power of the software is that it is quite good (but not perfect) at decoding CW, and with digital technology in an SDR and the algorithms in the software, you can 'hear' stations below the noise level. However, it isn't a replacement for your ears. CW Skimmer can be set to decode only CQing stations or anything it hears.

CW Skimmer is not a contest software package, but if set up correctly it could be used to aid an operator in a CW contest as it 'finds' many CW stations, decodes them and sends them to your cluster feature, which then populates the bandmap, thus helping operators find additional contacts. Naturally, the operator still has to go and work the stations.

The Great CW Skimmer debate

Whilst there are opponents who believe that this technology is the death-knell of radio, in past generations there have been those who said that about SSB when AM was commonplace, or when FM was introduced, or even when electricity was invented! You don't have to move with the times, and you can log with pen and paper if you want to. Yes, it presents more options, but it doesn't replace using your ears.

CW Skimmer is available from www. dxatlas.com/CwSkimmer/ (trial version) and costs US$75 for a full licence (at the time of publication).

Note that use of a skimmer in CQ Magazine contests (CQWW, CQWPX etc) puts you in to the Assisted category.

CW Decoders

Although not a contesting piece of software, there are pieces of software that help operators with poor CW skills to be able to 'listen' to (or realistically 'watch') and decode CW signals by transcribing what it 'heard' into plain text on screen. One package available is CWGet from www.dxsoft.com/en/products/ cwget/ ($35, 30€, unlimited free trial with limited features).

Basically, you connect the audio from the rig's output (using an audio cable/USB etc) in to your PC's line in/mic/USB socket and let the software decode what it hears in the passband, so if you use a 100hz filter you may hear only a couple of stations.

Advantages:

- Allows non-CW operators to operate CW by decoding CW signals for you into plain text
- Allows non-CW operators to work Search & Pounce by being able to decode the station they are trying to work over a series of QSOs with other stations, and then jump in to make the contact as you know what number/exchange to expect next. This makes CW less daunting and allows an unskilled

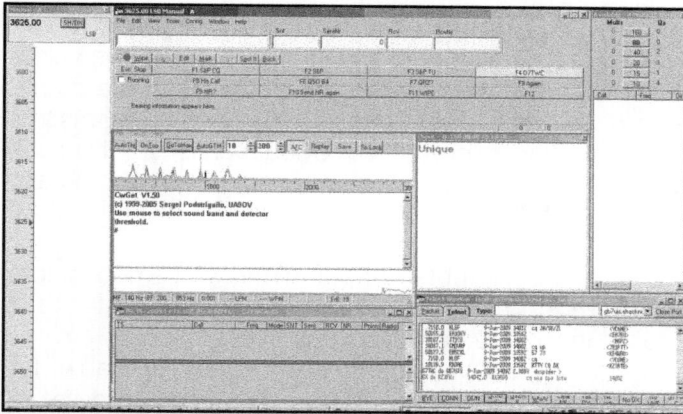

Fig 3.3: CWGet screen positioning

operator to take their time.

- Clear or strong signals can be decoded easily
- Works best, in the author's experience, when in the QRS Corral, i.e. the slow speed section of a contest band (if there is one), as stations coming back send relatively slowly, giving it more time to adjust the listening position of the software (see disadvantages, below).

Disadvantages:

- If you have a contesting package open, you need to also have CWGet open to view it. This may use valuable space on your screen, if you have a small screen such as a laptop.
- Signals need to be strong enough for the software to decode a signal. Ears are more sensitive in this scenario, so some experience of CW is always beneficial to fill in the gaps that the computer didn't hear. You need to position the red line at a level where decodes are consistent, but you also need to adjust it from time to time.
- When you click on CWGet it takes the computer's focus, so the software package isn't the PC's focus. If you want to then send something from the contesting package you first need to make it active, which involves an extra mouse

or key press, (e.g. ALT-TAB to toggle the focus in Windows) or click the mouse somewhere on the contesting software screen to make it 'active'. This can lose you valuable seconds.

- CWGet is not so good if you are CQing, as the stations don't reply on the frequency or pitch you are listening for. To cater for this there are AutoGTM (automatically go to the loudest station) or GoToMax (go to the loudest station when clicked) buttons. However, this works in conjunction with AFC functionality, but it takes a moment for the AFC to act, so you may lose the start of the callsign.
- If you don't use AutoGTM or GoToMax, or even AFC, then you need to click on the signal, which isn't instant because you need to hear or see the station before you can click on it. Once again you may miss some characters of the callsign.

Tip: If you position windows of the contesting software around the CWGet screen (as shown above), you may be able reduce the distance you need to move the mouse. It doesn't eliminate the focus issue but it does reduce it.

Satellite Tracking

For those interested in tracking the positions of satellites, various pieces of software can plot the course of the satellites in orbit. Various tools exist such as the free tool Gpredict (http://gpredict.oz9aec.net/) and Orbitron (http://www.softpedia.com/get/Others/Miscellaneous/Orbitron.shtml).

Amsat have a range of products advertised in their magazine. Kepler values to track satellite passes can be downloaded from various sources and are available on many phone apps too.

General Resources

A multitude of maps for CQ and ITU zones, grid squares, prefixes and other useful tools are freely available from www.dxatlas.com. A vast array of tools and articles are available from www.ac6v.com/software.htm and www.dxzone.com/catalog/Software/

Learning Morse

Useful tools exist such as Koch which incrementally builds up your morse alphabet in pairs of letters (free from https://www.g4fon.net/CW%20Trainer.php).

To practise CW in a contest environment, on the attached CD there is the excellent Morse Runner (http://www.dxatlas.com/MorseRunner/) or you can try G4FON's Contest Trainer (https://www.g4fon.net/Contest_Trainer.php). They both have adjustable speeds and replicate common macros used in contests, and have customisable noise and numbers of stations calling, just like in a real contest. You can start with a nice loud signal with no QRM or QSB and build up your confidence by working 'real' stations without fear of judgment or other stations nearby drowning you out. No radio is required, as it's all done through the sound card

Listing of a selection of packages

The listing below is a non-exhaustive selection of logging programs available. Other packages are available. Some programs that were present in previous versions of this book are no longer supported and therefore have been omitted.

N1MM+: https://n1mmwp.hamdocs.com (free)
SD: www.ei5di.com (free)
SDV (unsupported but still available, for VHF logging): www.ei5di.com/sd/sdvsetup.exe (free)
Writelog: www.writelog.com (subscription)
Win-Test: www.win-test.com (one-off payment)
DXlog: http://dxlog.net (free)
Logger32: www.logger32.net (free)
HamRadioDeluxe (HRD): www.hamradiodeluxe.com (subscription)
Fldigi: www.w1hkj.com (free), specialist package for data modes
Minos: https://minos.sourceforge.net (free)
WSJTX: https://physics.princeton.edu/pulsar/k1jt/wsjtx.html (free), specialist package for data modes incl. FT8 & FT4
SkookumLogger: https://www.k1gq.com/SkookumLogger (free), specialist package for Mac

Other packages are available.

4
Antenna Modelling

by Steve Nichols, G0KYA

Introduction

The average computer-literate radio amateur might initially ask why he should bother with antenna modelling, not being an antenna designer or guru.

However, every amateur is faced with limitations imposed by real estate, available supports, permission from their partner, neighbours or local council and will want to maximise the radiated signal from his QTH within these restrictions.

Antenna modelling using a home computer can answer many questions, particularly in comparing one possible antenna against another. It can also answer questions about the real performance available from commercial antennas, often described in glowing terms by manufacturers or suppliers.

All antenna modelling software likely to be of interest to the typical amateur is based around a modelling system called 'Numerical Electro- magnetic Code version 2' or NEC2. This was created in 1981 by the Livingstone Livermore Laboratories in California, the original client being the US Navy. Initially the system was classified, but over the years became available for general use.

Originally written in FORTRAN, the code has been translated over the years for use by the Microsoft Windows operating system.

NEC2 works by breaking the radiating elements of the antenna to be modelled into small portions called 'segments' and summing the overall electromagnetic radiation from the current and phase on these segments to produce the actual radiation pattern in a mathematical process called 'Method of Moments'.

Readers with a strong physics or maths background may like to peruse the original design methodology. It is available on the Internet at: www.nec2.org

A number of antenna modelling software packages are available based on NEC2. Some of these are freeware and some have to be purchased. It should be stated that the original NEC2 software is far from user friendly. It was written for professional antenna designers using mainframe computers and requires considerable additional add-ons to make it more intuitive and easier to use.

But products like EZNEC and MMANA-GAL are easy to use once the user gets their head around the process. Note that EZNEC and MMANA-GAL use different antenna design approaches, so if you learn one program you may find it difficult to move on to the other. Steve G0KYA wrote "An Introduction to Antenna Modelling" which uses the free MMANA-GAL software, so that might be a good place to start.

All versions define models on a 3-axis grid, X and Y orthogonally in the horizontal plane and along the Z axis vertically. Some knowledge of basic geometry (sine, cosine and tangent) and Pythagoras' Theorem can be useful to model with NEC2-based software, but neither are essential.

All antenna elements are defined as wires made up as a number of segments.

A more powerful version NEC5 is now available, with additional features beyond NEC2. This still has a restricted security status and requires the user to obtain a licence before purchasing the actual software as part of an antenna modelling package. Unfortunately, this costs several hundred pounds and would only be of interest to the serious antenna designer.

A list of available NEC-based antenna modelling software is detailed in **Table 4.1**.

It is left to the reader to select the appropriate package for themselves. They all do basically the same job and any of them will require an investment in time to get familiar with the features. Obviously, the software available as freeware or as free demo will be more attractive. Please note that Roy Lewallen W7EL has said that he will make EZNEC (PRO/2) free from the beginning of 2022.

Practical Example

We will now model a real life antenna, the popular G5RV multi-band dipole, complete with transmission line feeder to show how it is done. We will use the NEC2-based EZNEC program for this.

EZNEC is de-facto the amateur standard antenna modelling package and is relatively easy to use. EZNEC stands for Easy NEC! It is available as a download or via CD and runs on Windows 10 or Windows XP Service Pack 3 (SP3) or later, 32 or 64 bit.

EZNEC+ has 2000-segment capability, increased from the previous 1500. The EZNEC Pro limit has been increased from 20,000 to 45,000. Standard and Demo programs remain unchanged at 500 and 20 respectively.

A free download demonstration version with a maximum of 20 segments is available from the EZNEC website. This is also supplied with the ARRL Antenna Handbook on its accompanying CD. The free demo version can be used for the example about to be described, but the number of segments will be restricted to 19 rather than the 51 used. This results in inaccurate feed impedance and SWR values above 10 MHz. However, if this limitation is accepted, the radiation plots are similar.

Fig 4.1: EZNEC Desktop Logo.

The capability to model with more segments allows more complex models to be analysed but be aware that processing time increases with the number of segments used.

It is impossible in this short chapter to show all the bells and whistles available or all the caveats inherent in NEC2, but the hope is it will wet the appetite of the reader. A full 200-page manual is available from the EZNEC website.

Fig 4.1 shows the Desktop Icon for EZNEC (Version EZNEC V6.0). Double clicking on this produces the main EZNEC Control Centre screen – **Fig 4.2**.

This has a top menu, a left hand toolbar with but-tons and to the right of this another toolbar with buttons.

EZNEC does not allow you to start with a blank canvas, rather one of the pre-loaded example models needs to be opened and modified to suit the antenna to be modelled. After that, EZNEC always starts up with last antenna modelled. This is automatically saved under filename 'LAST.EZ' when the program is closed. The pre-loaded models enable the first time user to find what each menu item

Product Name	Supplier	Web Site	Free Demo Version?
MMANA-GAL	MM Hamsoft	https://hamsoft.ca/pages/mmana-gal.php	Freeware
4NEC2	Arie Voos	https://www.qsl.net/4nec2	Freeware
EZNEC V6.0	Roy Lewallen W7EL	https://www.eznec.com/	Yes
NEC2GO	Nova Plus Software	https://www.nec2go.com/	Yes
MININEC Pro	Black Cat Systems	https://www.blackcatsystems.com/	Yes

Table 4.1: NEC2-based antenna modelling software.

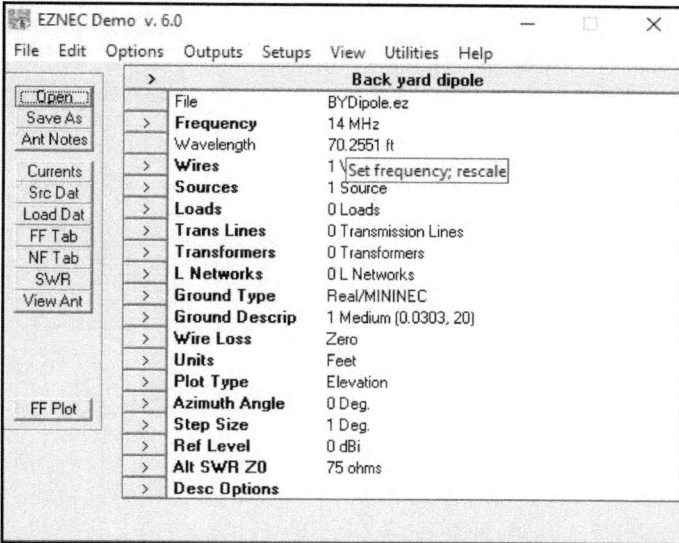

Fig 4.2: The EZNEC Control Centre Window with the sample model BYDipole opened.

does. Simple models of the basic antenna types, dipoles, yagis, and full wave loops are included in the preloaded library.

Click 'File' and then 'Open' and look for 'BY-Dipole' in the list of supplied example models. Left click on this, then click 'Open'. The Control Centre window reveals we have opened the model file BYDipole.EZ which is a 'Back yard Dipole' for 20 metres at a height of 30ft.

Click 'Wires', which opens the Wires definition window – **Fig 4.3**. The single wire dipole is defined as a wire with End 1 and End 2 xyz co-ordinates. The wire is 12 gauge bare

copper and is 33.43 feet long at a height of 30ft with 11 segments. The Wires window is where the antenna to be modelled is defined as a series of wires. We will be using only one wire, but more complex models will need many wires and it can be a tedious process to correctly define all of them. Certain tools are provided under the Create function to more easily define structures like radials, loops and helixes. Note the last two columns, dealing with insulation. This allows antennas with plastic coated wires to be modelled. The effect of the coating is to slightly shorten the required wire lengths compared to bare copper. We will now turn this 20m dipole into a 102ft G5RV dipole, made of copper wire, 0.1-inch diameter, complete with a 20m λ/2 feed stub made of 450-ohm open wire feeder. We will then examine the radiation plots and feed-point impedances at the bottom of the feeder.

Fig 4.4: The Wires Window edited for the G5RV. If the Free Demo version is used, the number of segments should be 19.

We have changed the length to 102 ft, the number of segments to 51 (19 if using the free demo version) and the wire diameter to 0.1-inch. To change these parameters, simply

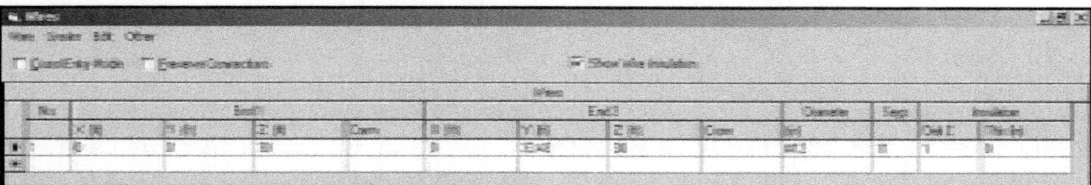

Fig 4.3: The Wires Window for BYDipole.

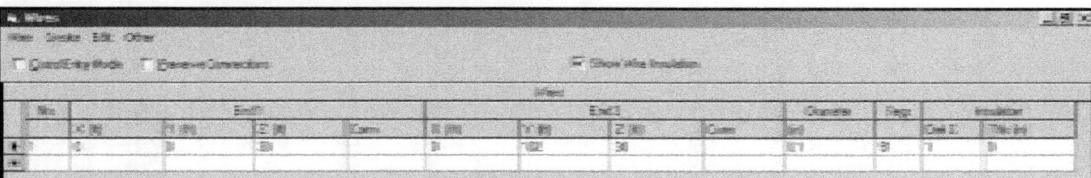

Fig 4.4 shows the Wires window, modified to create the model of the G5RV.

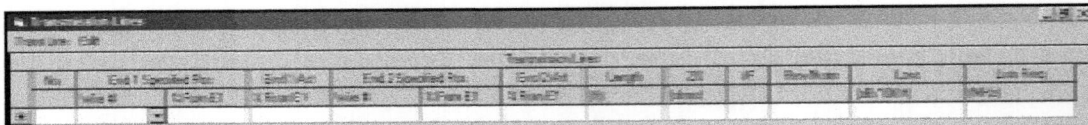

Fig 4.5: Blank Transmission Line window.

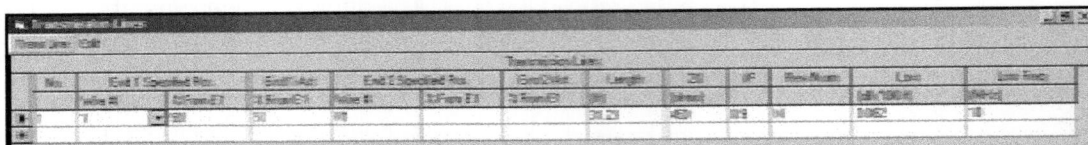

Fig 4.6: The Transmission Line Window completed for the G5RV.

highlight each value to be changed and type in the new value. Close the window when finished. The new values are automatically saved. **Note** we have used an odd number of segments. This is necessary with NEC2, to allow connection of feed points, or in this case transmission lines to the exact centre of the dipole.

We will now add a 20m λ/2 450-ohm ladder-line in the Transmission Lines Window. We need a half wave long line at 14.175MHz made from 450-ohm ladder line with a Velocity Factor of 0.9. This is 32.23 feet long. Real life transmission line losses are 0.082dB per 100ft at 10MHz.

Click on the Trans Lines button in the Control Centre window to open the blank Transmission Line window – **Fig 4.5**.

We now enter the required transmission line parameters. End 1 of the line is in the centre of Wire 1 at 50%. End 2 would normally be another wire defined in the Wires window. However, to make life simpler, we use an EZNEC specific shortcut and use a 'Virtual wire' (v1), which does not require a physical location to be defined.

Fig 4.6 shows the completed Transmission Line Window. Close the window.

Note that EZNEC V6.0 has a transmission line calculating engine built in, which will calculate the transmission losses for any given frequency and subtract these when calculating the antenna model gain.

Impedance transformations down the line are also calculated.

We now move the source from the centre of the wire to the virtual wire at the far end of the transmission line.

Open the Sources window. Change the Specified pos from wire 1 to v1 – **Fig 4.7**.

Click on OK to close the window.

We will now change a number of other model parameters to better reflect real life conditions

Click on the Ground Type button. In Real Ground Types, click on the High Accuracy button to change from MININEC ground - **Fig 4.8**. The High Accuracy ground model is more accurate than the MININEC ground. Click OK to close the window.

The High Accuracy ground model takes into account losses through the ground close to the antenna, the MIMINEC ground does not. The downside is wires can not be directly connected to ground with the High Accuracy ground making accurate modelling of vertical antennas more difficult.

Click on the Ground Descrip (Ground Description) button.

The Cond (Ground Conductivity) and Diel Constant in the sample model are set to those

Ground Type	Conductivity (Siemens/m)	Conductivity Milli-Siemens/m)	Permittivity
Poor	0.001	1	5
Average	0.005	5	13
Good	0.0303	30.3	20

Table 4.2: Typical ground parameters.

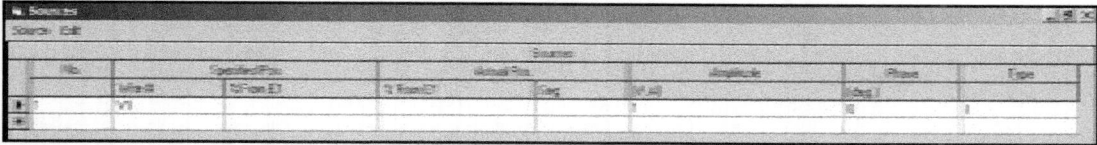

Fig 4.7: The Sources Window completed for the G5RV. Note we are using a 'Current' type source with amplitude of 1-amp.

for very good ground. To make the model more realistic we will change these parameters to those for average ground.

Change the Cond (Conductivity) value from 0.0303 to 0.005 Siemens per metre. Change the Diel Const (Dielectric Constant) value from 20 to 13 – **Fig 4.9**. Close the window.

This changes these values from those for good ground to average ground. These may not be those pertaining at any particular amateur QTH, the only certain way being to measure them.

However, an idea of the likely ground conductivity anywhere in the UK can be obtained from **Fig 4.10**.

Commonly accepted values for conductivity and permittivity values for good, average and

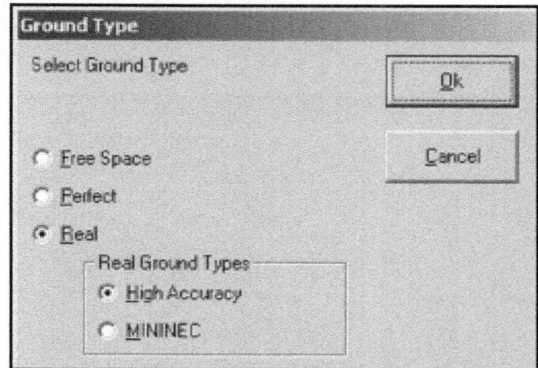

Fig 4.8: The Ground Type Window for the G5RV.

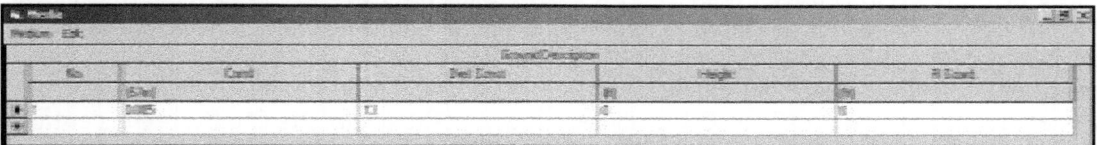

Fig 4.9: The Ground Description Window for the G5RV.

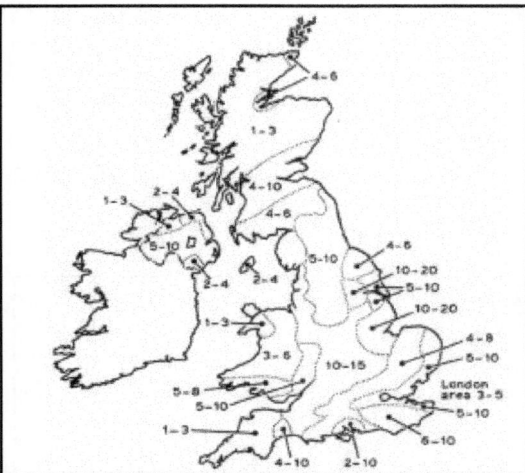

Fig 4.10: Ground Conductivity Map for the UK. The values are in mili-Siemens per metre.

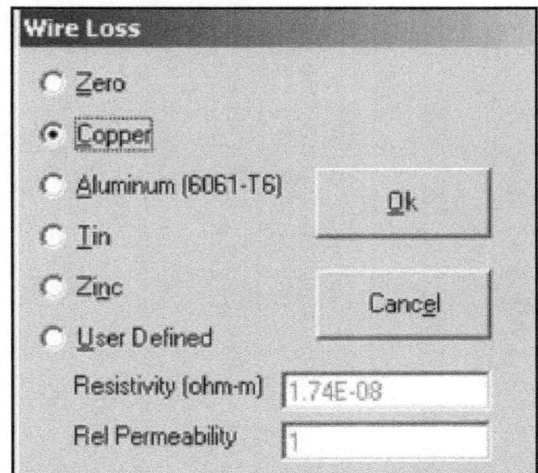

Fig 4.11: Wire Loss window for the G5RV.

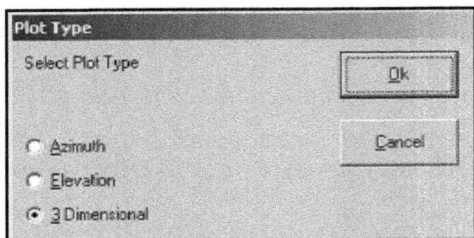

Fig 4.12: The Plot Type window for the G5RV.

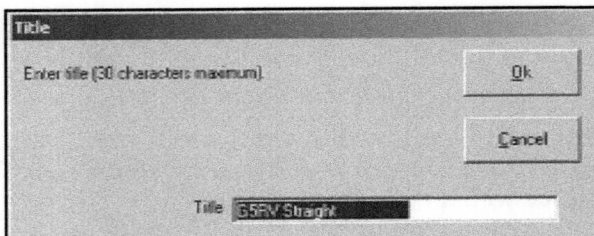

Fig 4.13: The Title window, changed for the G5RV.

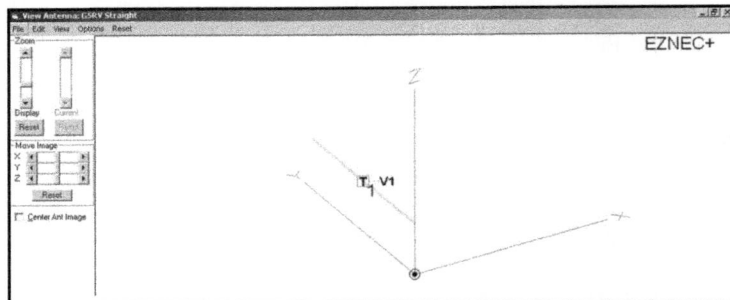

Fig 4.14: The View Ant window, showing the G5RV model.

poor ground are shown in **Table 4.2**.

Click on the Wire Loss button, click on the copper button to change from zero loss – **Fig 4.11**. Click OK to close the window. We are using copper wire and the resistive loss needs to be taken into account in calculating the actual antenna gain. Click OK to close the window.

Click on the Plot Type button and change from elevation to 3D by clicking on the 3D button – **Fig 4.12**. Click OK to close the window.

The 3D Plot type allows 3D as well as azimuth and elevation patterns to be displayed.

Finally, we rename the model G5RV and save to filename G5RV.

In the Control Centre window, click on the

description title tile at the top, (with 'Back yard dipole' in). In the change window, type 'G5RV Straight' - **Fig 4.13**, then click OK to close the window.

On left hand toolbar, click 'Save As', change the filename from 'BYDipole' to 'G5RV', then click 'Save'.

We are now in a position to model the SWR and feed impedances for each amateur band and plot the expected radiation patterns.

Click on 'View Ant' – **Fig 4.14**. This shows a simple representation of the physical antenna. Note the T, representing the transmission line and v1, representing the source or feedpoint at the end of the line. Note the controls on the left hand side, which allow the image to be zoomed and spatially manipulated. This can be useful in a multi-wire model, to see if all the wires connect correctly. Close the window.

Click on the Frequency button. Change the frequency to 14.175MHz, the exact centre frequency for the 20m band – **Fig 4.15** – then click OK to close the window.

At this point those using the free demo

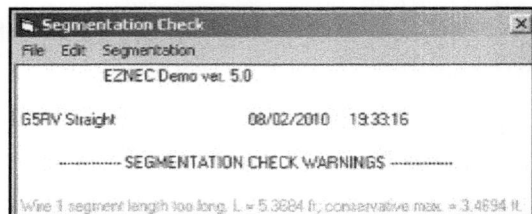

Fig 4.15: The Set Frequency window, set to 14.175MHz.

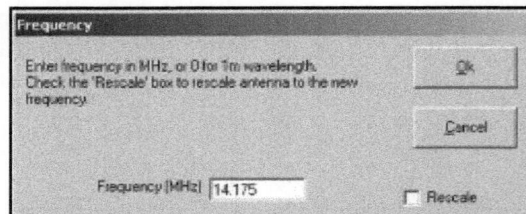

Fig 4.16: The Segmentation Check warning at 14MHz on the Demo version.

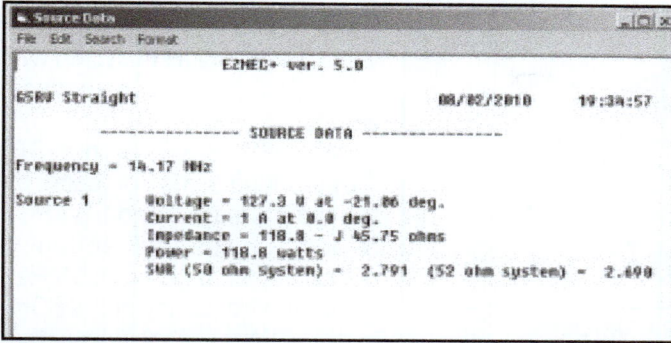

Fig 4.17: The Source Data Window, showing the feed Impedance for the G5RV model at 14.175MHz.

table showing the different values for impedance and SWR for the 19 and 51 segment models as calculated by EZNEC.

Click on the 'Src Dat' (Source Data) button. This window shows the feed-point impedance, SWR and other information for the G5RV at 14.175MHz. **Fig 4.17**.

Click on the 'SWR' button. Change the start frequency to 3.5MHz, the stop frequency to 30MHz and the Frequency Step to 0.5MHz. Click 'Run'. EZNEC now calculates the SWR at the feed point (end of the transmission line) from 3.5MHz to 30MHz in 0.5MHz steps and displays the results in a useful graph – **Fig 4.18**.

Note the dot pointer at 8MHz, which can be moved to each

version with a wire with only 19 segments instead of 51 will have got a Segmentation Check window, containing a Segmentation Check warning – **Fig 4.16**. This indicates that the segment is too long for NEC2/ EZNEC to calculate the correct imped-ance value. The warning indicates that the segment length of 5.3684 feet should be 3.34694 feet as a minimum. Close the window.

EZNEC will simply do a best effort when calculating impedance and SWR, but the values will differ from a correctly segment-ed model. The cure would normally be to increase the number of segments but there only 20 segments available in the free demo version. Later in the chapter I will produce a

frequency step. The impedance and SWR for the selected frequency are detailed in the bottom left hand side.

As can be seen, the SWR around many of the amateur bands is in fact quite high, not the 2.0 or less often expected. For the G5RV it is normal practice to connect 50-ohm coax at the end of the open wire feeder section to connect to the transceiver (ideally via a choke balun).

Transmission losses on the coax line, particularly at the high-er bands and with a longer length of coax, will be quite high. This can be reduced by using open wire feeder or ladder line all the way to the shack and a balanced antenna tuning unit, or placing a remote auto-matic antenna tuner (AATU) at the base of the ladder line. Either solution will result in lower transmission losses and is a better way of using the G5RV than the 'standard' method.

The SWR plot for any particu-lar amateur band can be looked at by changing the start and end frequencies, and by adjusting the

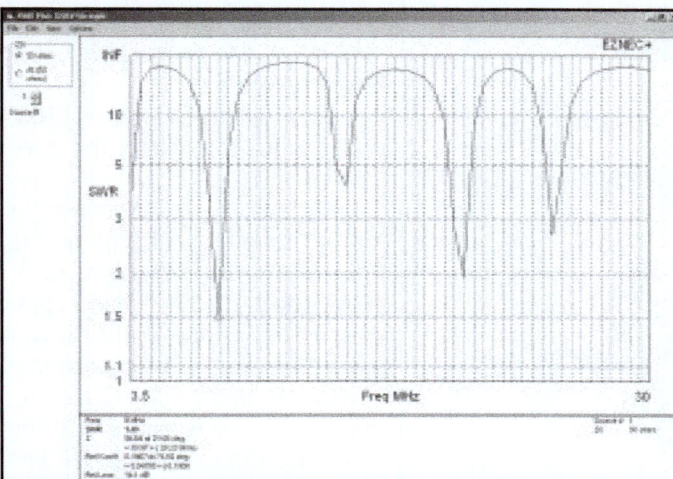

Fig 4.18: The SWR curve for the G5RV Model, from 3.5MHz to 30MHz.

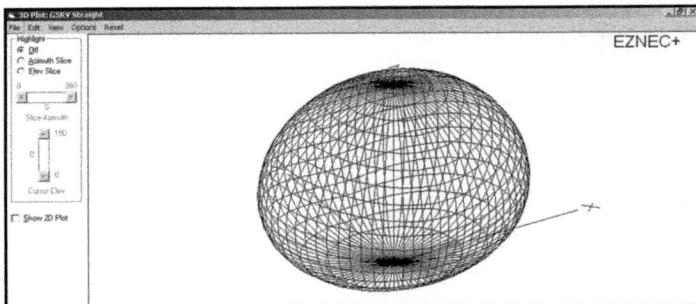

Fig 4.19: The 3D Far Field Plot at 3.65MHz.

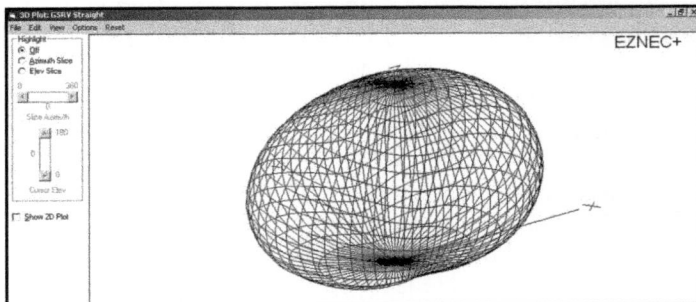

Fig 4.20: The 3D Far Field Plot for 7.1MHz.

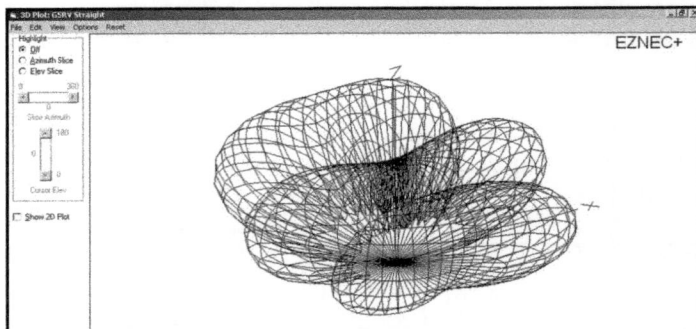

Fig 4.21: The 3D Far Field Plot for 14.175MHz.

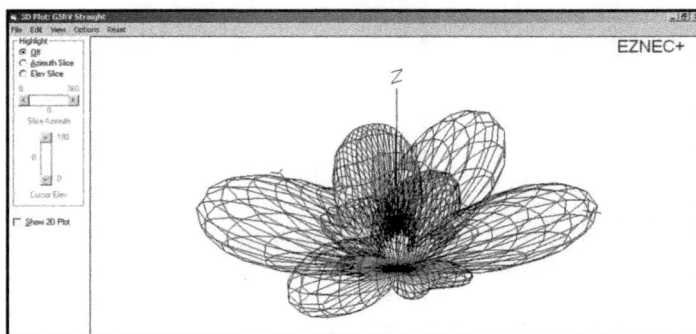

Fig 4.22: The 3D Far Field Plot for 21.225MHz.

frequency step to an appropriate value. Choosing a small step will take EZNEC longer to produce results.

We will now plot the 3D radiation patterns for each of the main amateur (non WARC) bands. Click the Frequency button and change to 3.65MHz, the centre frequency of the 80m band. Click 'OK' to close the windows.

Click the 'FF Plot' button. EZNEC now calculates the 3D radiation plot for 3.65MHz and displays it – **Fig 4.19**. Note that by clicking on the 'azimuth' or 'elevation' buttons then clicking the 'Show 2D' plot button we can change the view of the plot. Be careful with these 2D plots.

For the azimuth plot, the view depends on the elevation angle defined by the 'Cursor Elev' slider.

For the elevation plot, the view depends on the azimuth direction defined by the 'Slice Azimuth' slider.

The 3.65MHz 3D plot shows most of the transmitted signal will radiate at high elevation angles, due to the relatively low antenna height.

This is fine for local QSOs, but not for DX. Close the window.

Repeat the procedure for other bands by changing the frequency. Note that on 20M, 15M and 10M the radiation pattern is multi-lobe and of lower elevation angle. In the model the antenna is lined up along the y axis, assigning a real life compass direction to the y axis and using a Great Circle map will show the parts of the world

Freq (MHz)	19 Segments Complex Feed Impedance	19 Segment SWR	51 Segments Complex Feed Impedance	51 Segment SWR
3.65	18/54	6.2	18/56	6.5
7.1	63/-158	9.9	63/-147	8.9
14.175	120/-44	2.8	119/-44	2.8
21.225	54/142	9.3	55/177	13.3
28.5	2501/1051	58.8	2733/-556	56.9

Table 4.3: Differences between 19-segment and 51-segment modelling.

where the lobes will maximise DX performance.

The 3D plots for 7.1Mhz, 14.175 Mhz, 21.225 Mhz and 28.5 Mhz are shows in **Fig 4.20** to **Fig 4.23**. Note that at 28.5MHz a segmentation check on the 51 segment model will come up.

Table 4.3 shows the different impedance and SWR values derived fro the 19 segment and 51 segment models.

Finally in this example, we change the model from a straight dipole to an Inverted-V and show that the performance of this type of G5RV is vastly inferior to the straight version.

Fig 4.23: The 3D Far Field Plot for 28.5MHz.

We start with the saved G5RV file.

The Wire Windows for the inverted-V format is shown in **Fig 4.24**. Note that we have used three wires rather than two. There is a short wire in the centre (1.2 inches). This gets around a well-known problem in NEC-2, where wires meet at an angle. The ends of the Inverted-V are 3 feet above ground. In the Wires window, make the original wire 1 the short wire, but change the number of segments from 51 to 1. Add the two other wires, then close the window.

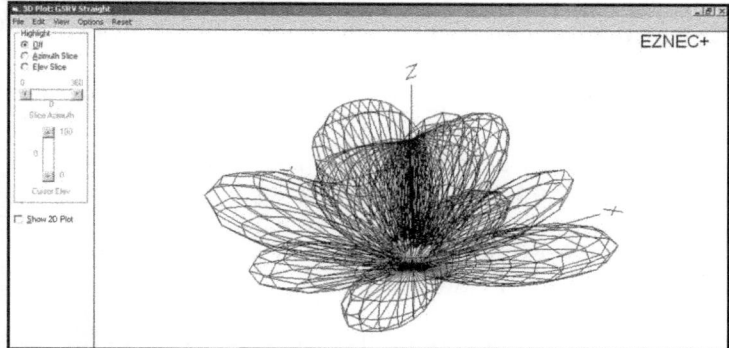

Change the description from 'G5RV Straight' to 'G5RV Inverted V' and Save as 'G5RV Inv V'. That's it!

The 'Show Ant' window for the Inverted-V is shown in **Fig 4.25**.

Look at the 3D plot at 14.175MHz – **Fig 4.26**. We now have a prominent lobe straight up, just right for picking up ionospheric noise and possibly unwanted short skip signals. Note also that the six lobes in the straight version have gone down to four. Use the Azimuth plot to show that the gain in these lobes is well down on the gain in the lobes in the straight model. Repeat for the 15m and 10m bands – it's the same.

Fig 4.24: The Wires window for the Inverted V G5RV Model

The moral here is to not install a G5RV as an inverted-V! Unfortu-nately, many amateurs will only have the one support, making the inverted-V format the only one possible. Further modelling will show better antenna solutions than the G5RV for these situations.

Conclusions

It has been shown how easy it is to model basic antennas using EZNEC.

In the process, some urban myths about the G5RV have been exposed.

From this it is hoped that readers will be encouraged into further modelling. It can be quite addictive, if you get into it! There is also the possibility of added interest, turning a model into reality and finding agreement between the model and physical reality.

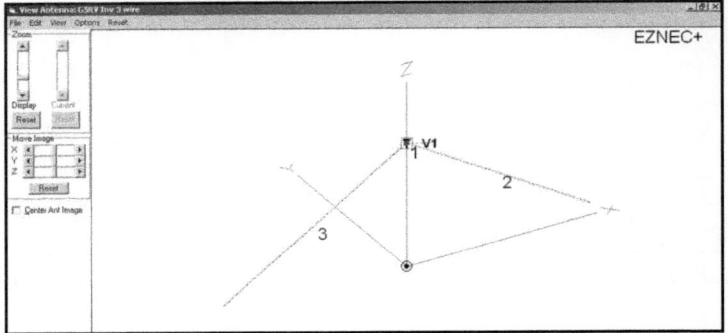

Fig 4.25: The View Ant Window for the Inverted V Model of the G5RV.

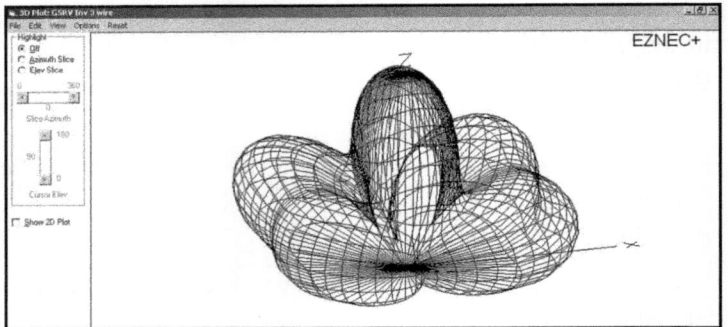

Fig 4.26: The 3D Far Field Plot at 14.125 Mhz for the Inverted V G5RV. The plot has been slightly rotated to better show the pattern. Note the prominent upwards bulge.

5
Propagation Modelling

by Gwyn Williams, G4FKH

This chapter will discuss how computers can be used in the shack to better the understanding of propagation. The topics will include information gathering, prediction production and beacon and ionospheric monitoring.

It would be frivolous to attempt a comprehensive list of places online where the various programs and information can be gathered. Instead, the keywords should be inserted into a favourite Internet search engine and this should be used as a basis for finding the required programs and interesting sites. One example of this would be to insert 'Realtime_foF2' into a search engine. One of the results will be 'Real Time foF2' from https://www.ngdc.noaa.gov. This will show a long list of stations that submit foF2 readings every 15 minutes. The resulting graphs show not only the current foF2, but the stations foF2 for the past 5 days.

It is of course possible to replicate most of what will be shown in this chapter without the use of a computer, assuming one has a scientific calculator and all the necessary algorithms to do the work and of course an awful lot of paper and even more time. I've been utilizing computers of one sort or another for over three decades to produce propagation predictions and now find the use of them indispensable. Let's continue with a discussion on the type of information required so that we are prepared for the various data inputs our computer programs require.

Internet Information Gathering

It is suggested that those interested register with such agencies as SEC NOAA SWPC and the Aus. IPS site. They will send bulletins each day which include the basic propagation indices. For prediction programs that utilize the ITURHFProp/VOACAP engine and derivatives, put 'predicted sunspot numbers' into your search engine for the sunspot numbers, use the predicted column AP and KP indices can be obtained in the recommended fashion. The Solar Terrestrial Dispatch site has a wealth of information and products aimed at the understanding of the current activity on the Sun and its subsequent interaction with Earth's ionosphere. SIDC in Brussels also has an interesting array of information and programs for users interested in propagation studies. Below are two other very important sites that are commonly used in propagation research: the Rutherford Appleton Laboratory (RAL), and the Solar and Heliospheric Observatory (SOHO) or SDAC on the NOAA site.

RAL is designated as a World Data Centre, so all sorts of archives and data can be found on the site. The section of most interest to those of us interested in radio conditions now or in the near future is the ionogram section. In this area ionogram pictures and ionogram data can be found. Ionogram explanations can also be found on the site. There was also a detailed explanation in the May 2009 edition of RadCom, the article has been reproduced and is available from the PSC section of the RSGB website. From ionogram data it is a relatively easy task to single-out a particular data type (e.g., foF2) and graph in order to better understand it. There is a section dealing with Ionograms

later in the chapter

An example of a simple line graph of the F2 layer Critical Frequency (foF2) is shown in **Fig 5.1**. This graph shows the diurnal effect of the foF2 at the bottom of the sunspot cycle and during the Spring period. The only real difference that will be seen at the height of the sunspot cycle will be the frequency magnitude. There are day-to-day and seasonal variations, for example during the winter months the foF2 tends to peak around the midday period. The Maximum Usable Frequency (MUF) at 3,000km can also be found within the same data set and will follow the foF2 graph because the two are interlinked. The SOHO site on the other hand contains a large amount of data shown mainly as images taken from the various satellites that have been put into orbit over the last several years. The site also contains a realistic visualization of the solar wind speed in the guise of a speedometer. Utilizing this information as well as that from the ACE RTSW site, it is possible to pinpoint the timings of such events as flaring and coronal hole occurrences. Both of these later phenomena adversely affect Earth's ionosphere and its ability to provide skywave communications.

Fig 5.2 shows one sunspot group 11017 in the North Western quadrant and a Plage region in the South Eastern quadrant. Plage regions are areas of the sun that have developed magnetically, but not substantially enough to be considered sunspot groups. Sunspot group 11017 belonged to the new solar cycle 24, whilst the Plage region belongs to the old cycle 23. This is determined by the polarity of the solar flux of the leading and trailing areas. For this solar cycle sunspot areas in the northern hemisphere have negative polarity leading with positive polarity areas trailing, and the reverse for the southern hemisphere. The polarity flips for each successive sunspot cycle. Occasionally during a solar cycle, a sunspot appears with a reverse polarity. This usually causes large-scale eruptions on the sun that can cause short-wave fadeouts, as well as widely dispersed aurora and Ground Levels Events (GLE's).

Propagation Prediction Programs

There are many different propagation prediction programs. It would be flippant to include them all, so we will concentrate on just two; ITUHFProp and VOACAP, both are gratis.

Before going on to do that it may be a good place to point out some research done into the accepted selection of prediction engine. In the August 2019 edition of RadCom, I wrote a short introduction to my Propagation Prediction Comparison Programs research. This compares the output of four prediction engines (ITURHFProp, VOACAP, REC533 and ICEPAC) with the recordings of the NCDXF beacon chain. This research is ongoing and This research is ongoing but here are some results so far: - It is hoped to increase the number of recordings to around 20,000 in the coming years. A final report will be submitted to RadCom when the analysis is complete.

The two engines chosen for further discussion are ITURHFProp and VOACAP, the first choice is obvious from the above graph and the second

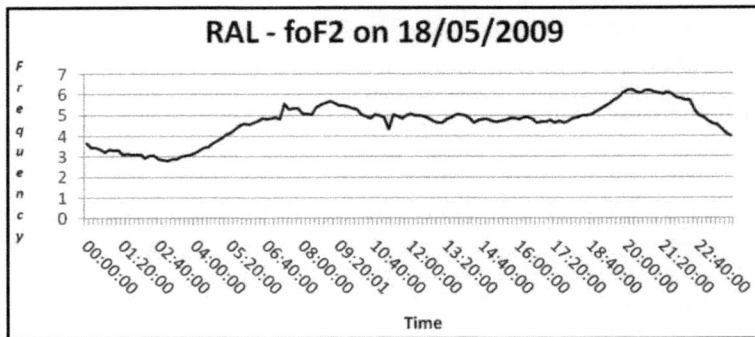

Fig. 5.1: A graph of RAL foF2 at the bottom of the sunspot cycle.

because it is so popular with Radio Amateurs and professionals alike.

ITURHFProp

The current version is 14.0 but 14.2 is due out sometime soon. As mentioned, it is freely available on the ITU web site under study group 3. It uses HF Model (P533) and the Noise Model (P372), both these papers are available through study group 3.

One of the reasons why this particular engine is not so popular as some others is because it does not have a GUI (General User Interface) however, there is a comprehensive guide to assist. I will not run through the input file,

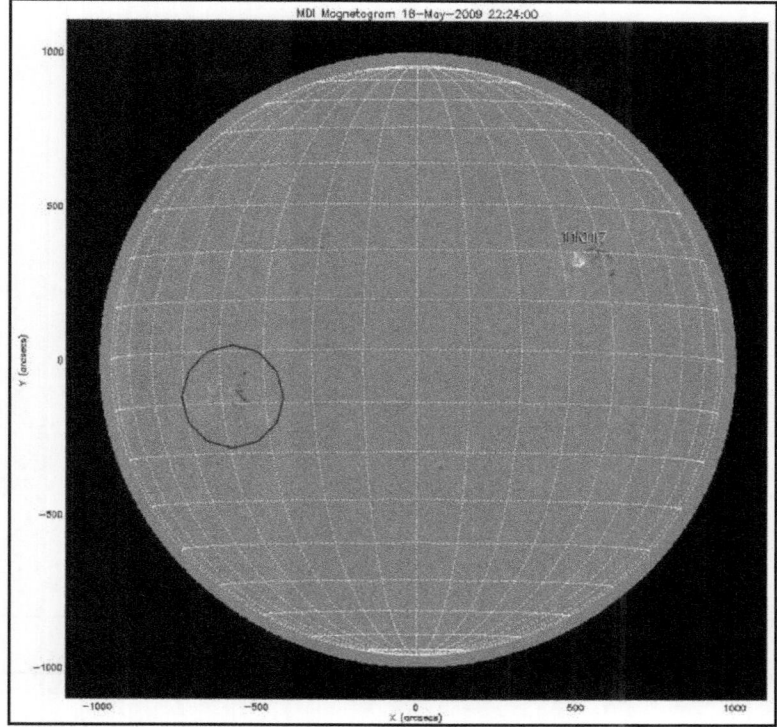

Fig. 5.2: Soho MDI image taken on 18 May 2009.

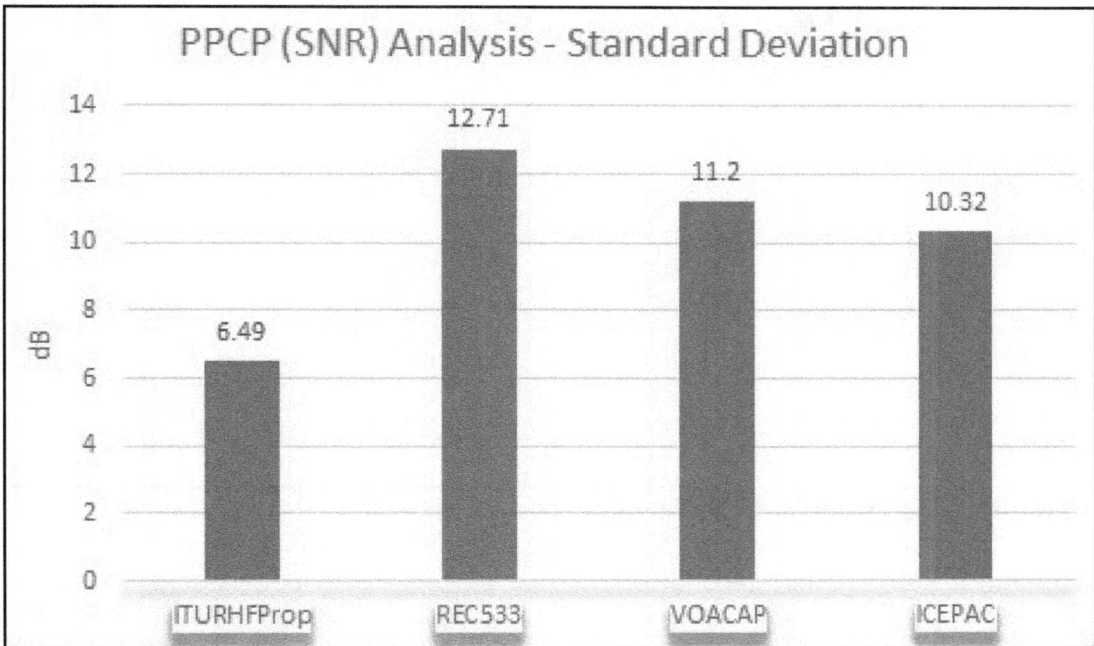

Fig 5.3 Depicting the standard deviation of 4,399 recordings.

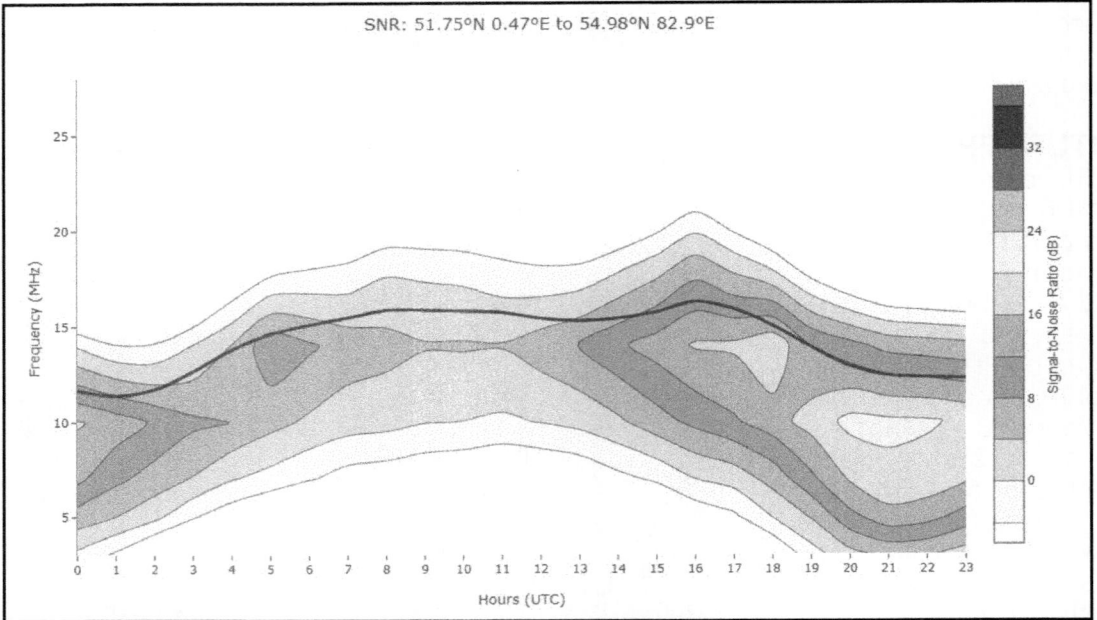

Fig 5.4. The MUF/SNR output graphic.

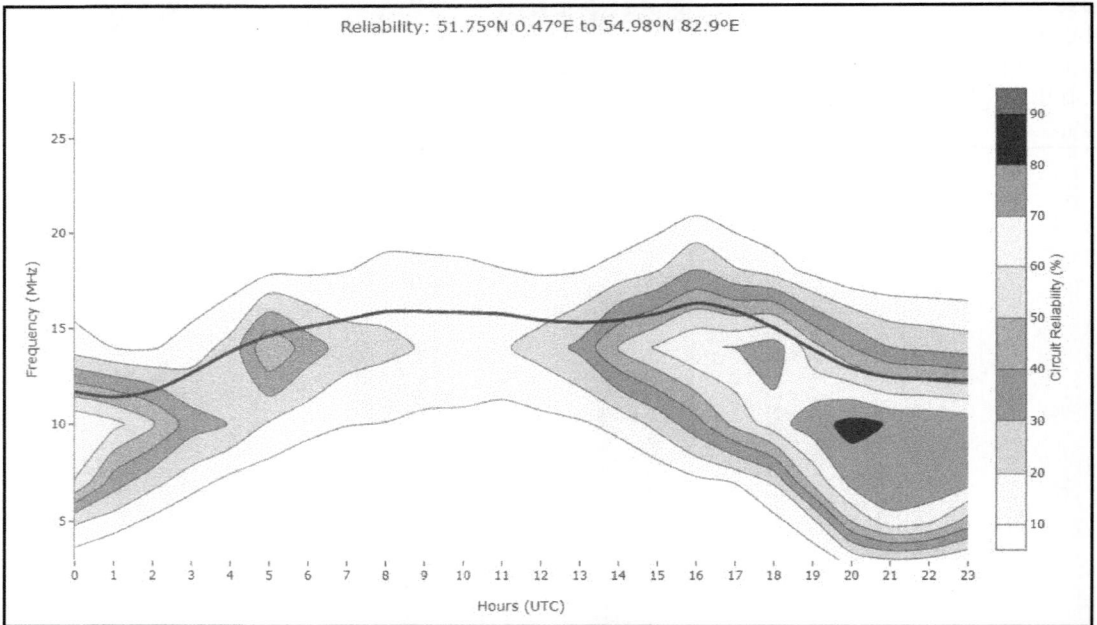

Fig 5.5 Showing the MUF and Reliability in percentage.

which incidentally is a text file and can be edited with e.g., Notepad. For ease, I have a web site; https://www.predtest.uk where not only is it possible to perform the predictions but to graphically view the output of various parameters such as SNR, Reliability and expected signal Magnitude. To prevent making this part of the discussion too long it will

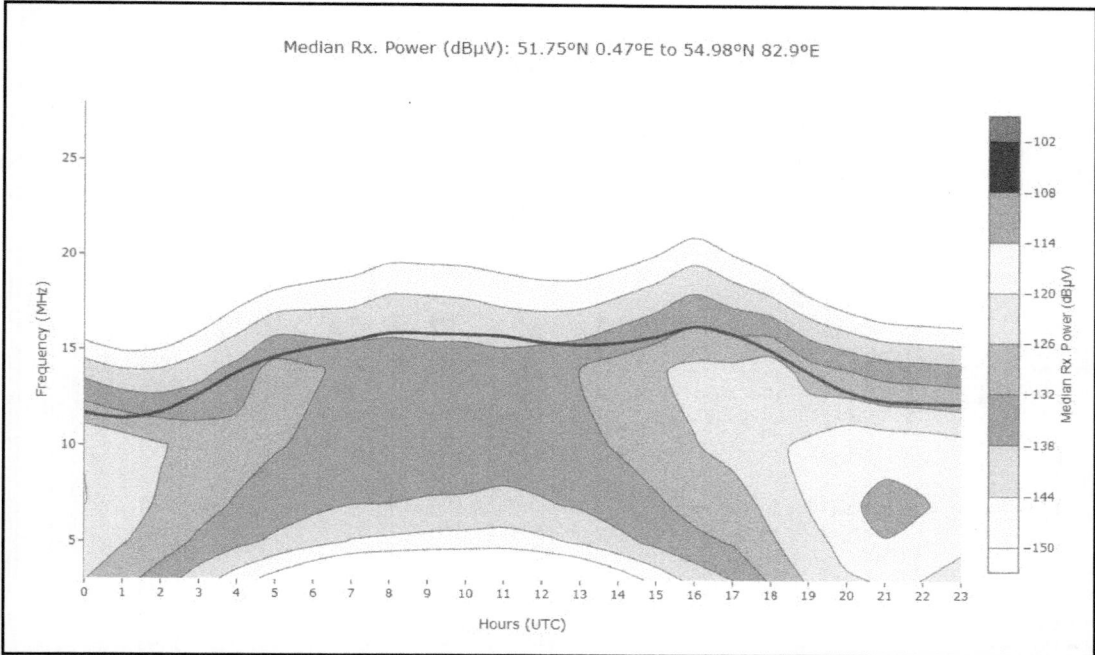

Median Rx. Power (dBμV): 51.75°N 0.47°E to 54.98°N 82.9°E

Fig 5.6 MUF and Median Rx. Power.

concentrate on a Point-to-Point prediction. It is the third choice along on the top ribbon, the first two concern area coverage. There are two ways in which one may input the Tx. and Rx. locations, (i) using the upper map, moving the pins changes those parameters in the lower form. The other way (ii) is to put that information into the form manually. Choosing the required date allows the program to automatically look up the SSN so there is no requirement to know it. Choose both the Tx. and Rx. aerials simply complete the rest of the System Requirements, clicking on 'Generate Predictions' brings up the analysis screens. The first output graphic presented is for the MUF and SNR.

The MUF (maximum usable frequency) is the purple line, the heat map shows the SNR (signal to noise) levels. Hovering the mouse over the graph on the web site shows the data for the time and frequency. The next choice along on the output banner is MUF and Reliability, shows below:

As with the previous graphic there is a MUF line and a heat graphic, hovering the

mouse shows the relevant data. The next output choice shows the MUF and Median Rx. Power (signal magnitude):

Similar to the previous two outputs only this time showing the magnitude of the expected received signal strength. A trick that some miss is to scroll down past the above graphic to show a further graphic which depicts the magnitude in S-Point form. The last output graphic choice is the text output for this prediction.

The other main prediction categories work in a similar fashion, so the site should be relatively easy to manipulate and produce one's own predictions, the output graphics can be down loaded by hovering the mouse on the graphic and clicking the download button on the right above the graphic.

VOACAP

This can be downloaded from the internet, put 'Greg Hand VAOCAP' into a search engine and one of the choices will be 'HF Prediction Models...', click on that and you

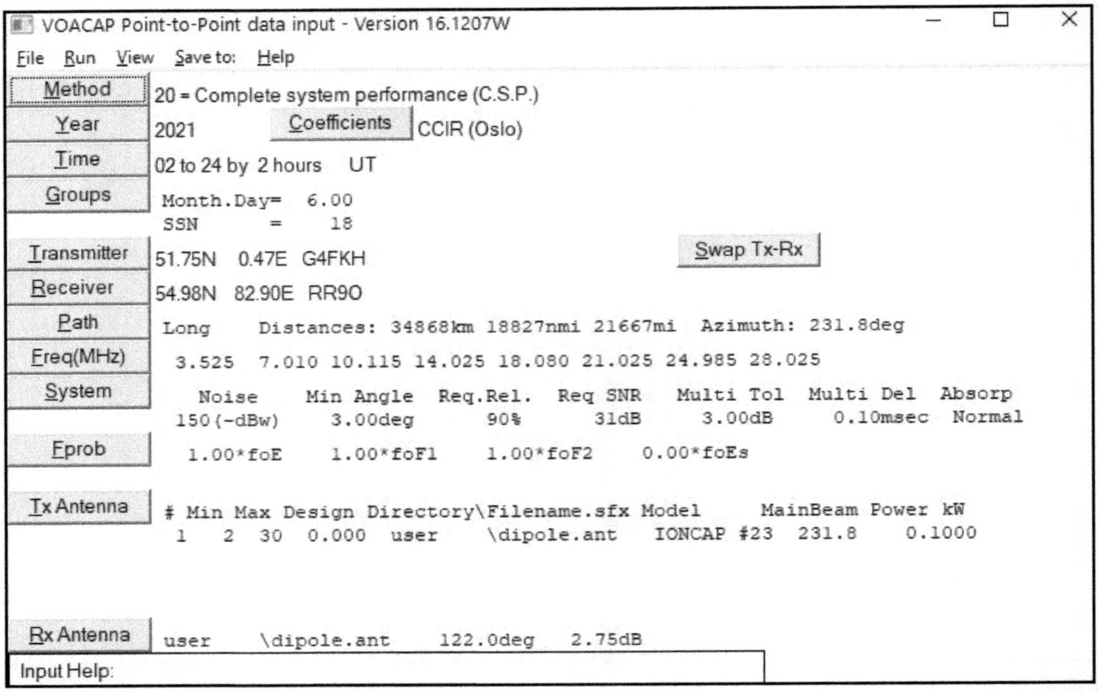

Fig 5.7. VOACAP input screen.

will see a download list. I do not know how often (if ever) this suite is updated and if it is, whether the noise data is updated at the same time, perhaps this is why it fares so badly in my analysis so far. There are many guides on the internet for specific data output such as SNR etc. but for this exercise we will go through a Point-to-Point prediction, the input screen:

Most of the fields on this form are self-explanatory, one has to know the required SSN of course. Clicking on the field on the left-hand side brings up the various relevant choices. Leave the 'Fprob' alone unless you really know what you're doing. When choosing an aerial, if it is required to have them pointing at each other (radio wise) click the 'at RX' and 'at TX' buttons on the aerial pop up. Click the 'Run' button and a small pop-up will show the various ways to manipulate the engine, choosing 'Circuit' will produce a textual output, a snap-shot of one is below:

This is only a part of one hour's output, there are many more parameters depicted.

However, for our purposes, look at the SNR output for 14.0MHz, which says 31dB, this should produce a loud CW signal and even a good SSB signal. The graphical SNR output:

The graphic ties in nicely with the textual output at 06:00, an SNR of 31dB. All the output parameters may be viewed in a similar manner by clicking on 'Parameters' on the top banner.

My research so far has shown VOACAP to be very poor at least during the lower part of a solar cycle for sort distance predictions, i.e., those below about 3,000Km. But must better than the others when over say about 9,000 Km. So far others programs have not been considered because most of them are derivatives of VOACAP, however, there are exceptions; Proplab-Pro and the Australian GRAFEX naming two. Both of these are quite expensive and as far as I can tell do not lend themselves to output manipulation.

Lastly, I would like to mention the RadCom predictions which recently added data mode predictions. In the September 2020

```
Jun    2021            SSN =  18.               Minimum Angle= 3.000 degrees
  G4FKH                RR9O               AZIMUTHS          N. MI.
KM
  51.75 N    0.47 E - 54.98 N   82.90 E      51.80  302.03    2784.1
5155.8
  XMTR   2-30 IONCAP #23[user\dipole.ant      ] Az= 51.8 OFFaz=  0.0
0.100kW
  RCVR   2-30 +  5.2 dBi[samples\SAMPLE.00     ] Az=302.0 OFFaz=  0.0
  3 MHz NOISE = -150.0 dBW      REQ. REL = 90%     REQ. SNR = 31.0 dB
  MULTIPATH POWER TOLERANCE =   3.0 dB   MULTIPATH DELAY TOLERANCE =   0.100
ms


   6.0 15.4   3.5   7.0 10.1 14.0 18.1 21.0 25.0 28.0   0.0   0.0   0.0 FREQ
       2F2   4 E   4F1  3F2  2F2  2F2  2F2  2F2  2F2    -     -     -  MODE
       9.9   5.0 18.8 17.8  8.6 11.0 11.0 11.0 11.0    -     -     -  TANGLE
      18.4 17.5 18.9 19.0 18.2 18.5 18.5 18.5 18.5    -     -     -  DELAY
       373   90  262  351  340  402  402  402  402    -     -     -  V HITE
      0.50 1.00 0.94 0.84 0.75 0.06 0.00 0.00 0.00    -     -     -  MUFday
       161  359  199  167  156  224  343  479  479    -     -     -  LOSS
       -15 -226  -60  -25  -11  -77 -195 -329 -328    -     -     -  DBU
      -141 -339 -179 -147 -136 -204 -323 -459 -459    -     -     -  S DBW
      -169 -152 -160 -164 -168 -171 -173 -175 -177    -     -     -  N DBW
        28 -187  -19   16   31  -33 -150 -283 -282    -     -     -  SNR
```

Fig 5.8a. Part output from VOACAP

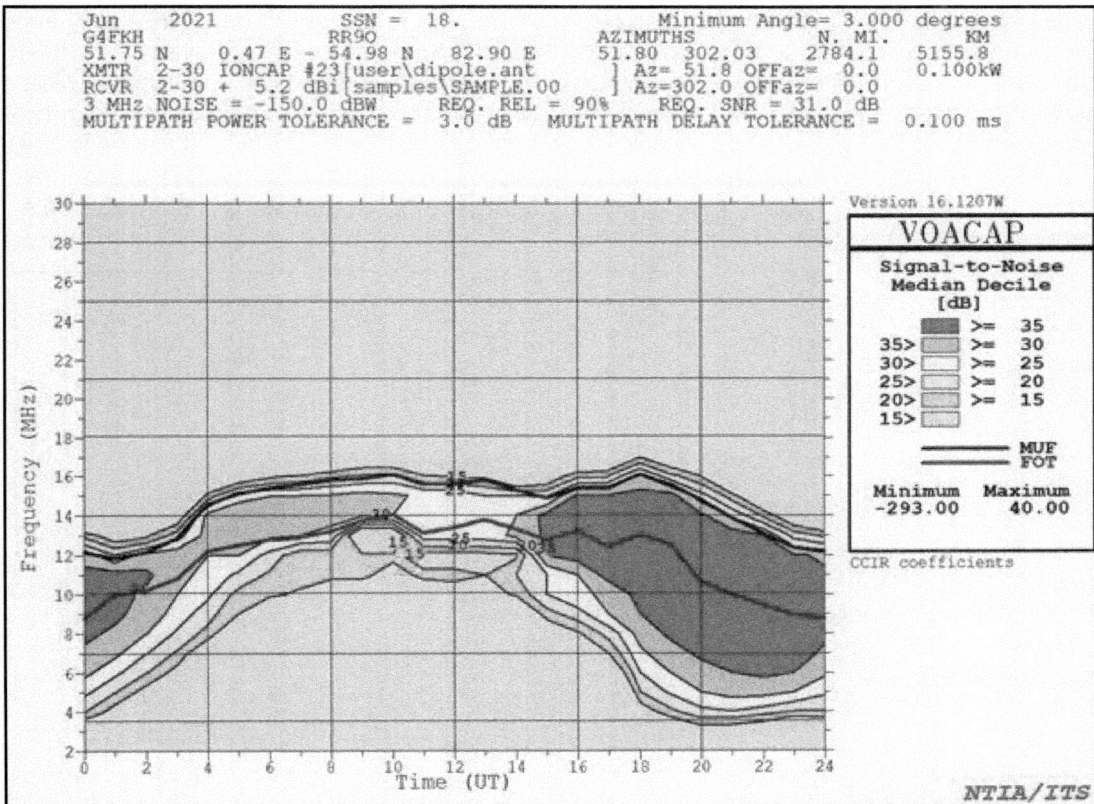

Fig 5.8b. The same information as the previous text output but in graphical form.

edition of RadCom, Peter Duffett-Smith G3XJE and I jointly published an article on this subject. There is a calculation method to produce data mode predictions from the RadCom table. I have found the simplest way to use this is to put it into a Spreadsheet and just change the Figs as necessary.

Beacon Monitoring Programs

This next set of programs use disparate methods to assist in the under- standing of propagation.

In this section we will talk about beacons and tools for monitoring them. These tools, with the exception perhaps of Spectrum Lab, also show a definite slant towards HF. Beacons can assist us to ascertain what part of the world is open for communication and of course more importantly, when. The NCDXF/ IARU Beacon Chain Network is perhaps the most important group of beacons for the HF bands. Established by the North Carolina DX Foundation (NCDXF), in cooperation with the IARU, it operates a worldwide network of high- frequency radio beacons on 14.100, 18.110, 21.150, 24.930, and 28.200MHz. Beacons are necessarily stable in two ways. Firstly, they should be stable in frequency and secondly, they should transmit on a timely basis. Monitoring programs have differing ways of keeping an accurate track of time and this will be explained as the various programs are examined. It would, therefore, be logical to start this section with a look at what programs cater specifically for the NCDXF beacon network.

FAROS

This program was written by Alex Shovkoplyas, VE3NEA in 2006, but unfortunately it is not gratis. However, it is the best program that the author has come across for monitoring the NCDXF beacon chain, whether free or not. This is primarily because of the way it differentiates between the CW ident and QRM, which unfortunately is quite common on the protected frequencies.

The NCDXF beacon network has eighteen beacons that utilize the five frequencies once every three minutes. For that reason, timing is paramount. Faros handles this by monitoring a number of time servers on the Internet. Consequently, a broadband connection is necessary if accurate results are required. The program will communicate with a modern transceiver, changing frequency as required. There is very little in the way of setup, just the normal questions to answer, such as home Latitude/Longitude, COM port and type of audio card. The program then talks to the transceiver and logs the beacons when they are heard. It has a very clever algorithm built in which very satisfactorily identifies the beacons from the background noise and inter- ference and logs the information in a text file.

All the beacons are recorded, whether or

```
;DATE=2009-06-02
;UTC-      -MHz-   -Call-   -SNR,dB-QSB,%-Evidence-Delay,ms
13:15:00    14     4U1UN     -6.2     100     0.20       50
13:15:10    14     VE8AT      7.3       2     4.25       34
13:15:20    14     W6WX      -7.5     100     0.18       61
13:15:30    14     KH6WO     -7.4     100     0.10        1
13:15:40    14     ZL6B      -5.6     100     0.38      227
13:15:50    14     VK6RBP    -6.0     100     0.24      335
13:16:00    14     JA2IGY    -6.4     100     0.50      242
13:16:10    14     RR9O      -7.1     100     0.10      -18
13:16:20    14     VR2B     -14.2     100     0.12      165
13:16:30    14     4S7B      -6.0     100     0.09      258
13:16:40    14     ZS6DN     -4.9     100     0.26      182
13:16:50    14     5Z4B      -4.5     100     0.48       43
13:17:00    14     4X6TU    -17.5     100     0.15      189
13:17:10    14     OH2B      11.5       0     6.77       25
13:17:20    14     CS3B      -5.5     100     0.28       41
13:17:30    14     LU4AA    -20.0     100     0.05      -43
13:17:40    14     OA4B      -7.4     100     0.22      -53
13:17:50    14     YV5B      -4.9     100     0.16      -51
```

Fig 5.9: A sample of Faros' output text file.

										Time															
QRG	Beacon	00	01	02	03	04	05	06	07	08	09	10	11	12	13	14	15	16	17	18	19	20	21	22	23
14	4X6TU							0						0	0	1	2	0	0	1					
14	5Z4B									0						0	1		0						
14	CS3B						1																		
14	KH6WO							0																	
14	OH2B							3	2	3	1	1	1	1	1	1	1	2	2	2					
14	RR9O													1	1	1	1	0	0	2					
14	VE8AT							1	2	1	1	1	1	1	0	0	0	0	0	0					
14	YV5B																0								
14	ZS6DN																		0						
											Time															
QRG	Beacon	00	01	02	03	04	05	06	07	08	09	10	11	12	13	14	15	16	17	18	19	20	21	22	23
18	4X6TU									0				1	0	0	0		0	2					
18	5Z4B														0										
18	OH2B							1	2	2	1	1				0									
18	ZS6DN										0														
											Time															
QRG	Beacon	00	01	02	03	04	05	06	07	08	09	10	11	12	13	14	15	16	17	18	19	20	21	22	23
21	4X6TU										0	0			0	0									
21	CS3B												0												
21	OH2B							0	0				0		0										
											Time															
QRG	Beacon	00	01	02	03	04	05	06	07	08	09	10	11	12	13	14	15	16	17	18	19	20	21	22	23
24	4X6TU									0	2	1	2	1											
24	OH2B									2	4	2			2										
											Time															
QRG	Beacon	00	01	02	03	04	05	06	07	08	09	10	11	12	13	14	15	16	17	18	19	20	21	22	23
28	4X6TU									0	1	1													
28	CS3B											0													
28	OH2B										2	2													

Faros - NCDXF Beacon Chain Monitoring - May 2009 by G4FKH

Table 5.1: Faros computed output in a spreadsheet format. The Figs represent average S-point numbers at the various times.

not they are audible. For post reception interpretation a method of analysis is required. Whether it is with an Excel spreadsheet or external program becomes the user's choice.

Going back to the log file information it is only when the SNR, dB is above 0 that signals are actually heard, also the QSB should be much less than 100% and the evidence should probably be around 3 or more. During the period shown and, on the frequency shown, only two beacons were heard; VE8AT and OH2B. The delay in milliseconds is also of use. With this parameter it is possible to work out when the path is via long-path (LP) as regards the more normal short-path (SP). I've used this parameter for an extensive study of the ZL6B beacon which comes in via LP during the early mornings from about November to about March, the study can be found on the NCDXF web site. There are other displays available but they do not render down too well in black and white. It is possible to download the pro- gram for a trial if required. An important application that the

author has written uses the output log from this program to prepare an Excel spreadsheet of the output as shown in **Table 5.1**.

The procedure to arrive at the spreadsheet output is quite involved, but we are discussing the use of computers in the shack. The Faros output is firstly run through a PERL program that decides what beacon reports are strong enough to be audible. The output from the PERL program is in an Excel format, so it is a simple procedure to then add the file to an Access Database (the author has a 22,569-record database going back to April 2006). It is then necessary to run an

Access 'Crosstab' query that puts the output in roughly the format shown above. All that is then necessary is to smarten it up a little so that it looks like the spreadsheet above. Now, what practical use has this spreadsheet? Looking at it closely, it sort of resembles a RadCom HF Propagation Predictions page, this output is used to ensure that the HF Propagation Predictions for RadCom are in the correct ball park. A lot of the bea-

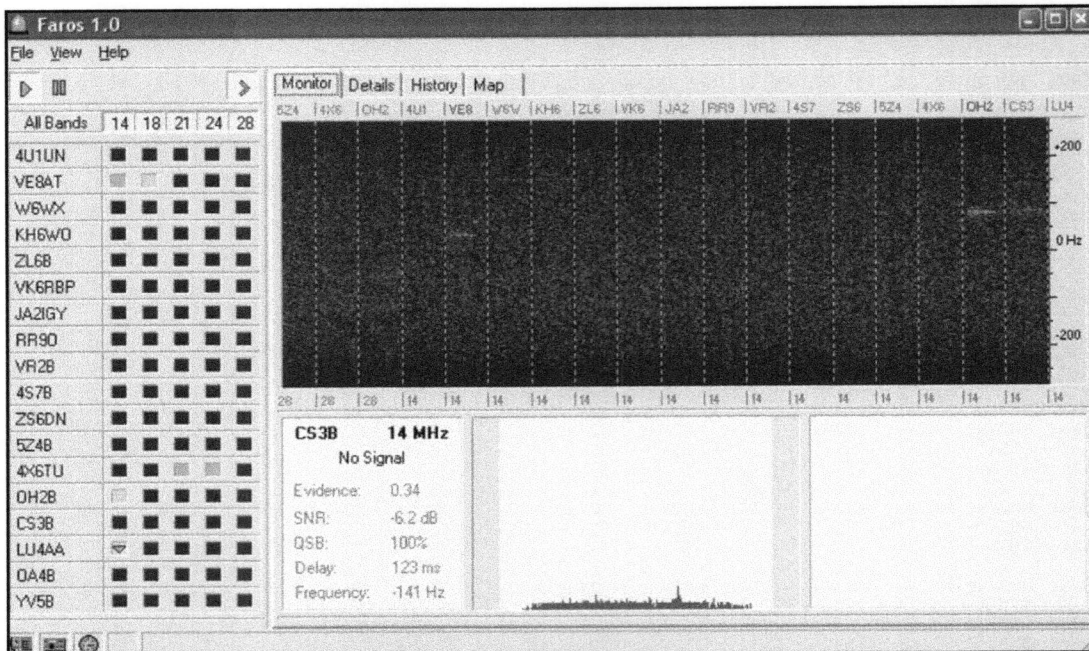

Fig 5.10: The Faros Monitor screen showing which beacons were heard on which bands in the last 15 minutes. The scrolling larger display on the right highlights the last 15 beacons heard.

cons are in the same general vicinity as the destinations listed in RadCom so they lend themselves very well to this purpose.

Spectrum Lab

Spectrum Lab is a much more complicated and far more sophisticated tool than Faros, but it can be utilized to monitor most anything including beacons. The main consideration is therefore whether the PC internal clock is correct, if using an older PC, older than a Vista operating system, this program as well as a lot of others require accurate timing. So they should be used in conjunction with a GPS receiver. These receivers can be found occasionally, quite cheaply on eBay. Spectrum Lab is the brainchild of Wolfgang Buescher, DL4YHF, and is basically a Spectrum Laboratory that can be used for applications that are in the audio part of the spectrum up to those in the GHz bands and all frequencies in between. It also has an RDF capability. Built in application files

include radio equipment tests, slow Morse reception, and Digi modes. However, the first-time user will be required to carefully read the instruction manual and make the necessary changes to the setup. This program has a great deal of promise but there is a very steep learning curve involved. The rewards, however, will be great for those able to comprehend the intricacies.

Bespeak

BeSpeak5 is one of those programs that show when a particular NCDXF beacon is transmitting. Others are available from the NCDXF website. There is a decent manual and automatic logging feature with this program that quite accurately differentiates between QRM and the required beacon signal. It does this by discriminating between background noise and Morse. Written by Alan Messenger, G0TLK, it has been developed over the years and has 2D and 3D graphical output. Alan responded quickly when

Fig. 5.11: Bespeak 5 main screen, showing all controls and buttons.

an update was requested for the program, and it enabled a more in-depth discussion here. The entire application is controlled from the primary screen. This and the two other screens used for beacon monitoring are shown in **Fig 5.9**, **Fig 5.10** and **Fig 5.11**.

The three screen shots are used for monitoring purposes.

Once the setup is completed it is only necessary to select the monitoring band from the initial screen. With the program control- ling the receiver, all the necessary commands are sent to it, making changing bands a one-button clicking action. There is a comprehensive help facility with this program and it is necessary to ensure that it has been read, under- stood and all the appropriate actions carried out prior to proceeding with monitoring operations. The graphical output is quite good, but does not render down with sufficient accuracy to be displayed here. There is no facility with this program to allow post logging analysis, so it is suggested that those interested download

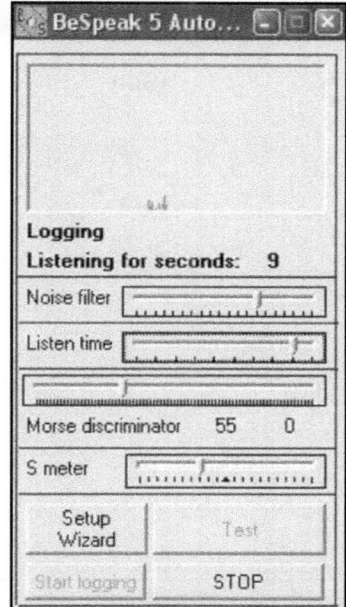

Fig 5.12: Morse discrimina-tor screen and its controls.

it and try the built-in analysis, after all the program is gratis.

GB3RAL

This is the name that the program originator, Peter Martinez, G3PLX, has selected for this application. It provides a graphical and textual output when logging the three 5MHz beacons; GB3RAL, GB3WES and GB3ORK on 5290MHz (GB3ORK is the only beacon currently transmitting). These beacons were put in place by the RSGB's 5MHz working group to assist with the understanding of propagation on that frequency and specifical- ly, to ascertain the extent of NVIS propagation for inter-G working. There is a help system and again it is necessary to read all of the available information and to set the program

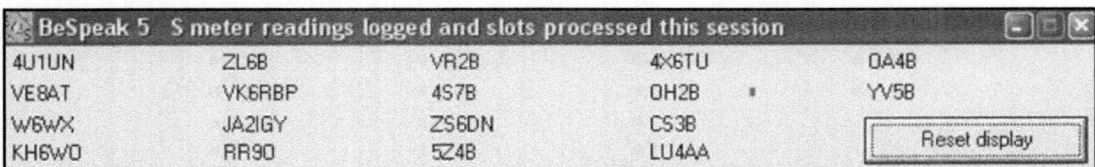

Fig. 5.13: Logging screen. NB: OH2B was logged with an S-meter reading of 1

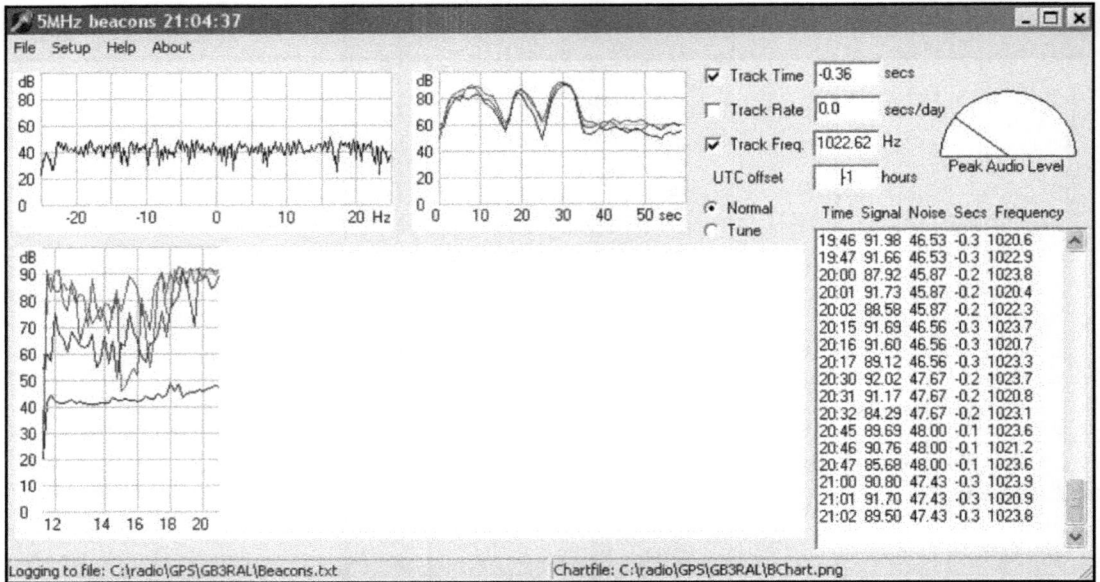

Fig 5.14: GB3RAL's main and monitoring screen.

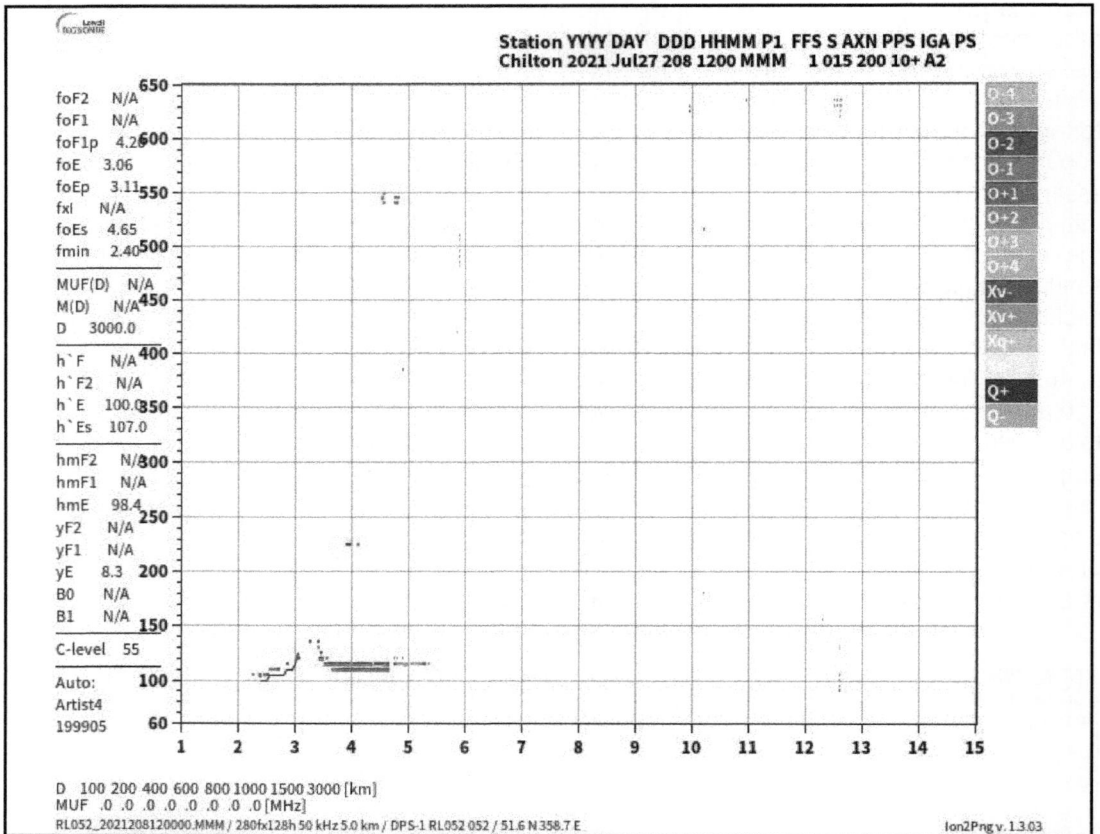

Fig 5.15. RAL Ionosonde showing blanketing Sporadic-E.

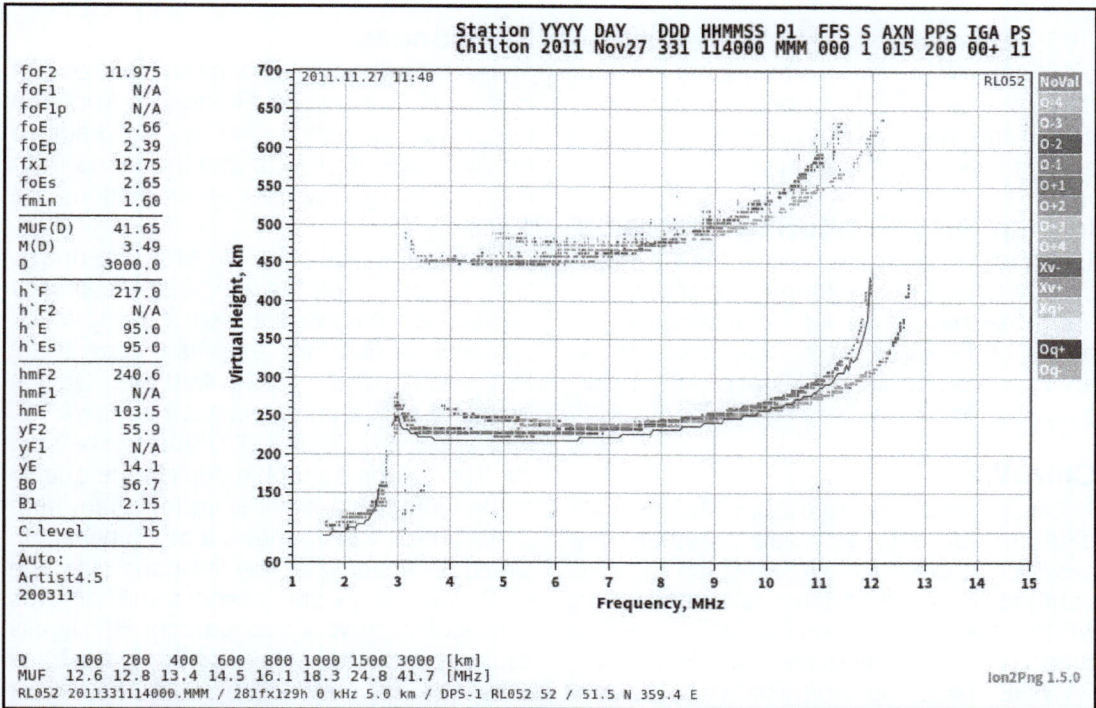

Fig 5.16. A RAL ionogram from back in 2011.

up as described. Once this has been completed, logging can proceed with S-meter reading of 1. the minimum of user input. The main screen is depicted in **Fig 5.15**.

The screen shot looks a little busy at first glance, but when you become accustomed to it all the information becomes meaningful. Starting at the top left is the FFT display. When transmitting, beacons should show a peak around the '0' point. The next on the right is the dB output against time of the last three beacons. To the right of these are setup and information displays. The help section should be consulted if clarification is required. The bottom right is a running record of textual output for the recorded beacons. The bottom left is a graphical representation of the three beacons received over time.

Fig 5.14 shows 10 hours of recording. The very top line at the left-hand side in the graph is the GB3RAL beacon. The next is the GB3WES beacon and the third is the GB3ORK beacon. The lowest line represents the noise

level in the receiver. All this information and more are recorded on disk in a text file. Fig 5.14 shows some interesting trends. For example, the GB3RAL signal started to fall off about 1200. This is because the foF2 dropped. When it recovered around 1800 the signal for that beacon was strong again.

This can be directly attributed to nearly NVIS working, while the other two beacons at the receive location are received via the E-layer. In the very early days of the GB3RAL beacon, when analysis was first being performed, it was discovered that GB3RAL was being heard late at night when the foF2 was far below the transmission frequency. To assist with analysis, ionogram pictures were downloaded from the RAL web site and it was discovered that the extra-ordinary wave was responsible for the reception. This was the first time that the author realized that the extra-ordinary wave was as useful for NVIS, as it is at HF frequencies (as will be shown later). Another peculiarity noticed was that the

GB3RAL program was showing the reception of signals when they were inaudible, thus they were very weak. It may be possible to exploit this peculiarity by utilizing the more exotic modes using computers.

Ionospheric Monitoring Programs

This category of programs includes those that can be used for monitoring the ionosphere and those that can be used to understand the nature of the ionosphere. Also included are applications and datasets that provide basic concepts and measured figures.

ChirpView

These are two separate applications from different sources, but used together they provide an excellent way in which the ionosphere can be monitored. ChirpView was written by Andrew Senior, G0TJZ, whilst Stepper was produced by Arend Harteveld, PA1ARE. These are sophisticated packages that come with a pre-requisite of a GPS receiver for accurate timing. A good knowledge of using computers with amateur radio is really another pre-requisite for this pair, as the setup is not at all simple.

ChirpView is a program for receiving signals from ionospheric chirp sounders – transmitters that radiate an unmodulated carrier which is swept at a constant rate across the HF spectrum in order to study ionospheric propagation conditions. These sounders are perhaps unfamiliar to many radio amateurs, but can be thought of as all-band beacons. To receive these signals properly requires a special receiver that tracks the transmitter along its sweep.

Such hardware is impractical for most amateurs who are more likely to have an HF receiver with SSB capability. Using such a receiver, a passing chirp sounder signal produces a short chirp in the receiver of around 30 milliseconds duration. ChirpView can detect these chirps and measure their timing precisely to enable propagation to be monitored on the frequency to which the receiver is tuned.

Ionosondes

There are two sources of ionosonde output in the U.K. one is at the Rutherford Appleton Laboratory and the other at RAF Fairford (really USAF). Below is an Ionosonde from RAL, showing how poor HF conditions are presently:

These graphics are quite simple to understand once you get the hang of it. As already mentioned, I have written a short article which appeared in RadCom, there is a copy in the PSC section of the RSGB web site. On the left-hand side parameters such as the foF2, foEs and the MUF at 3,000Kms can be read off. The coloured graphic depicts the quality of the ionosphere. Here is an Ionogram from 2011 which clearly shows a much better HF situation: Reading off the left-hand side, the foF2 was 10.725MHz and the MUF was 35.89MHz, so very strong upper HF signals could be expected. The slightly fainter trace above the main one is the second bounce of the ionosonde signal. This is explained a little further in the article. The extraordinary trace in the one shown in green.

The beauty of these ionograms is that for instance, if the HF bands are not performing to the level suggested by any prediction engine, then sometimes a quick look will show why. It may be because of Blanketing Sporadic-E or just low foF2 and MUF levels. Blanketing Sporadic-E blocks out the HF bands so basically very little can get through. It does however provide enhancement to VHF signals.

Field Strength

This is an ITU database of medium skywave field strength in dB above 1uV/m, normalized for 1kW EIRP for a total of 181 longpath and shortpath circuits. This database can be used to ascertain when a circuit should be available without the need to consult a propagation prediction program. The database covers two full sunspot cycles. It is necessary to apply to the ITU for free downloads, which are limited to three per annum, otherwise a payment of around 20 Swiss Francs is necessary.

US/UK World Magnetic Model -- Epoch 2005.0
Main Field Horizontal Intensity (H)

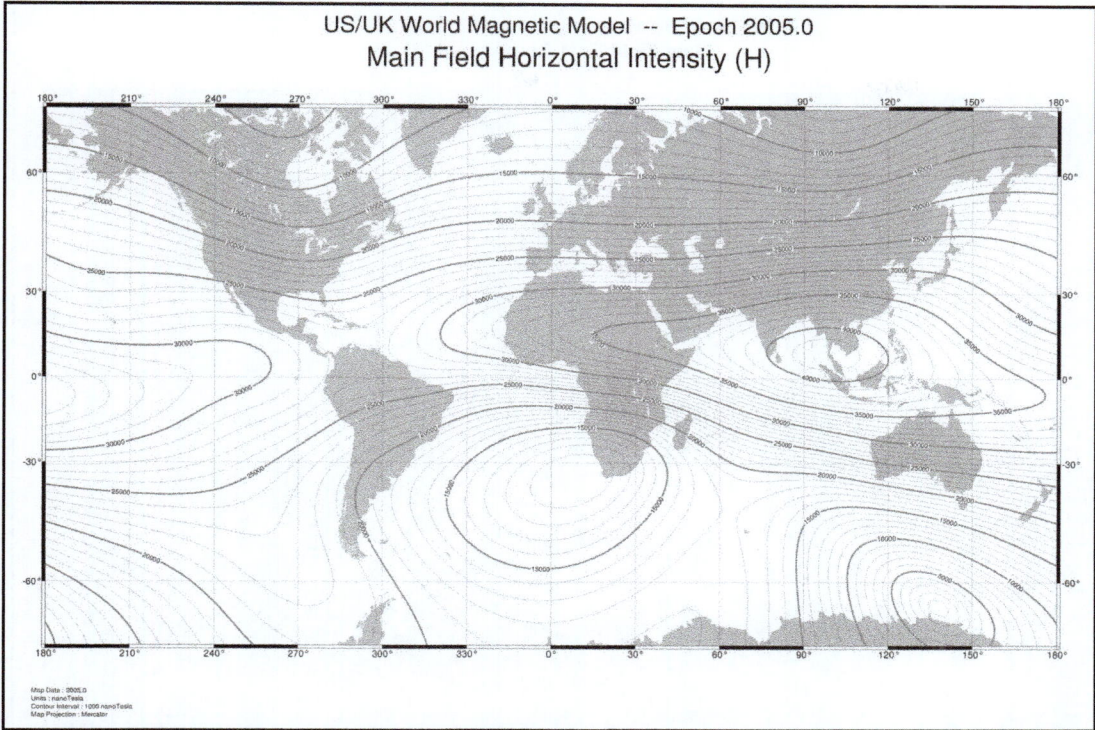

Fig.5.17: The world's horizontal magnetic field.

Noise

There is a set of programs available under the above scheme for download from the ITU site. It provides characteristics and applications of atmospheric radio noise data and man-made radio noise. There are three programs in the suite: NOIS1 and NOIS2 and NOISBW. The first two give values of atmospheric noise, man-made noise and galactic noise from Recommendation ITU-R PI.372. The only difference between these two programs is the style of the output. The latter program provides all the parameters relating to atmospheric noise.

GeoMag

Inserting 'NGDC GeoMag' into a WWW search engine gives the URLs for various programs associated with the earth's Geomagnetic state. An interesting world map of the horizontal component is shown in **Fig 5.17**.

This clearly shows, among other things,

that the magnetic equator does not coincide with the geographic equator; it is to the North by roughly 10 to 15 degrees. **Fig 5.18** is of the East component and assists in the visualization of why North/South paths are generally, much superior to East/West paths. Radio signals contain a magnetic property which, to put it very simply, resists crossing these contours.

Proplab Pro

The last piece of software in this section and indeed for the chapter is a propagation laboratory. Produced by the Solar Terrestrial Dispatch; they claim that it is the most advanced radio propagation ray tracing system in the world. It is certainly the most advanced that the author has seen for radio amateurs. It is a state-of-the-art soft- ware package not for the feint hearted. It is also a little expensive.

The current version is a lot simpler than previous offerings. The reasons for this are primarily because the majority of the background

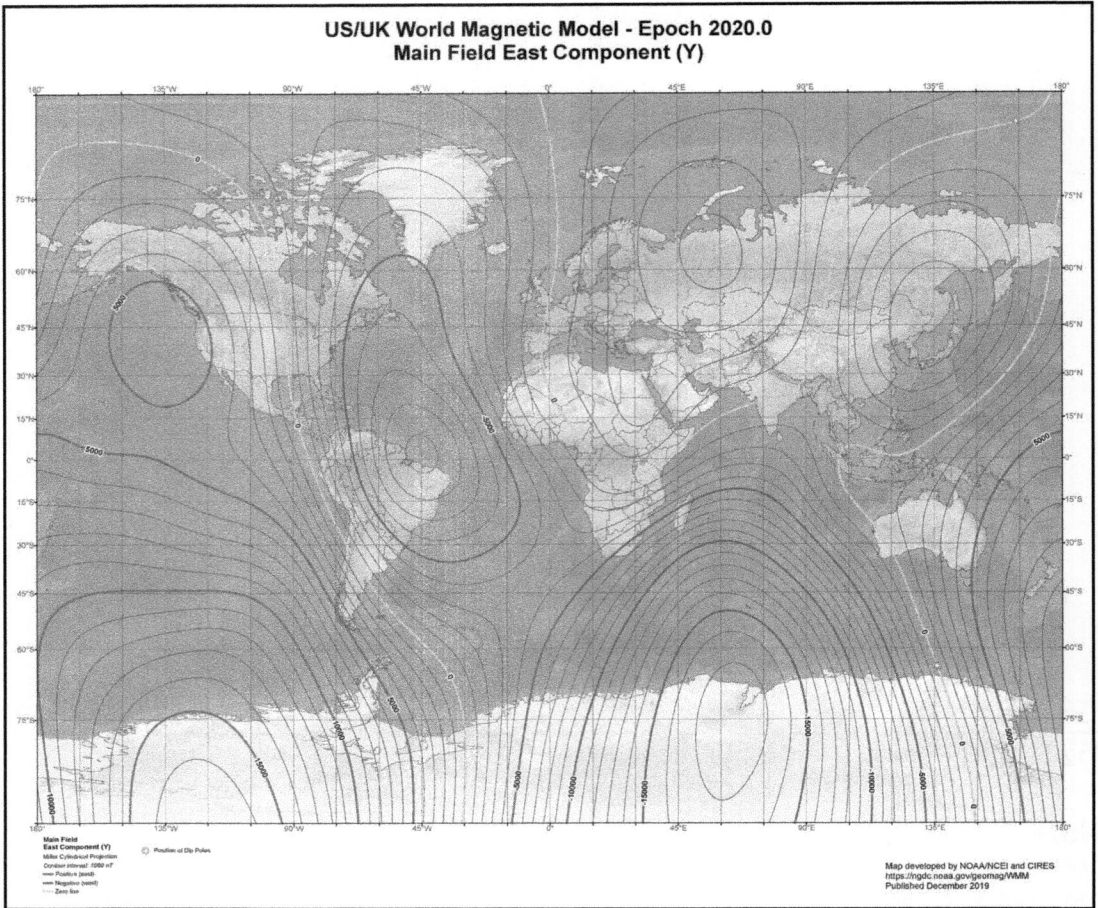

Fig 5.18: Earth's main field East component.

input is now gathered by the program off the Internet at the click of the mouse, plus it's a lot more complex 'under the hood'. Ray tracing can be performed in either 2-D or 3-D, the two illustrations – **Fig 5.19** and **Fig 5.20** detail a ray trace from the author's location to 3B8 (Mauritius). Another output from this package that I particularly like is the world MUF display, shown in **Fig 5.21**.

The contours show the varying MUFs over the globe, but there are a multitude of other parameters available to display. The circuit described is plotted by a great circle line between the two locations along with the day/night terminator line. Hop lengths are known, so it is quite easy to see the frequency in the

mid each separate one. The lowest frequency from these hops should be about the best for the circuit. This quick analysis is not always accurate and it is for this reason and others that ray tracing has been included in the package. Learning More About Ionospheric Propagation

To fully understand propagation, it is necessary to have a firm back- ground in mathematics and basic ionospheric propagation theory. The programs mentioned previously will then become clearer and more useful as a consequence.

How does one educate one's self in propagation? Well, there are several ways. One is to trawl the Internet, looking for sites that explain all about propagation. A good example is the

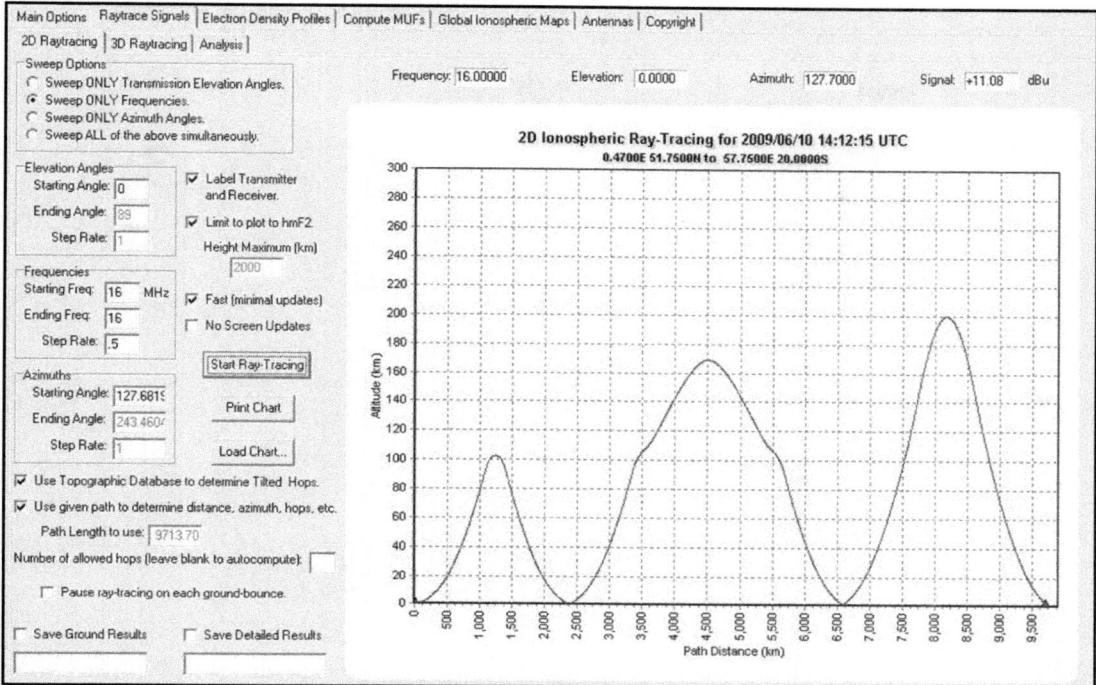

Fig 5.19: 2-D ray tracing output of Proplab-Pro. The start point is on the extreme left and the destination is the little triangle at the right.

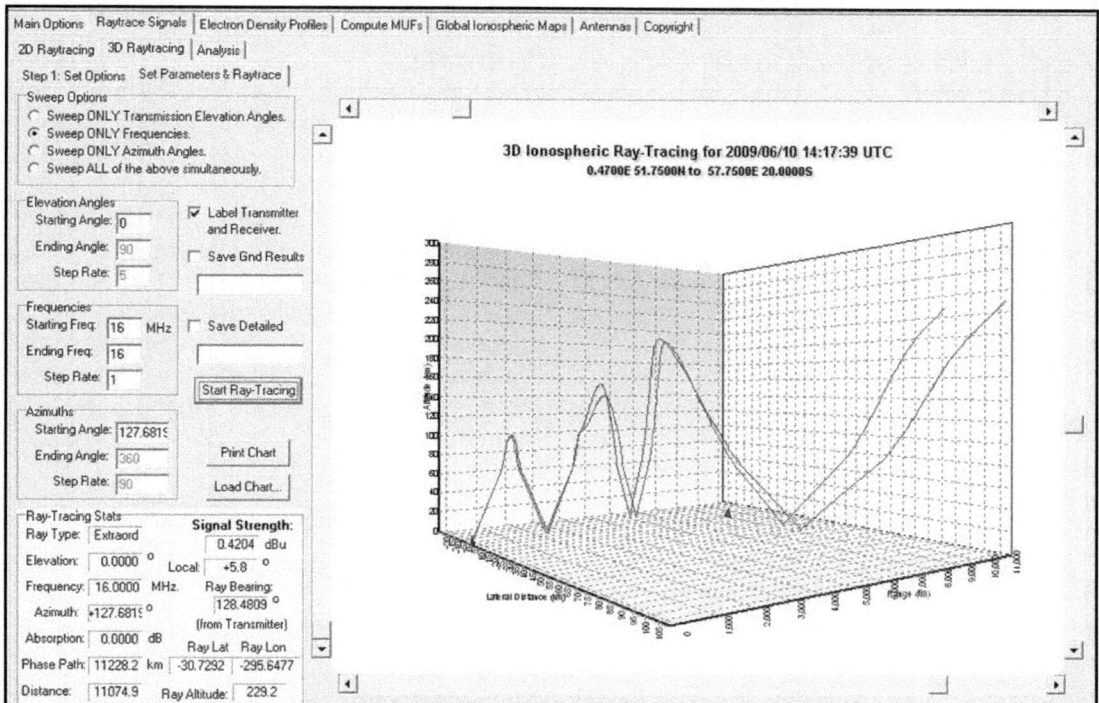

Fig 5.20: 3-D ray tracing output for the same circuit.

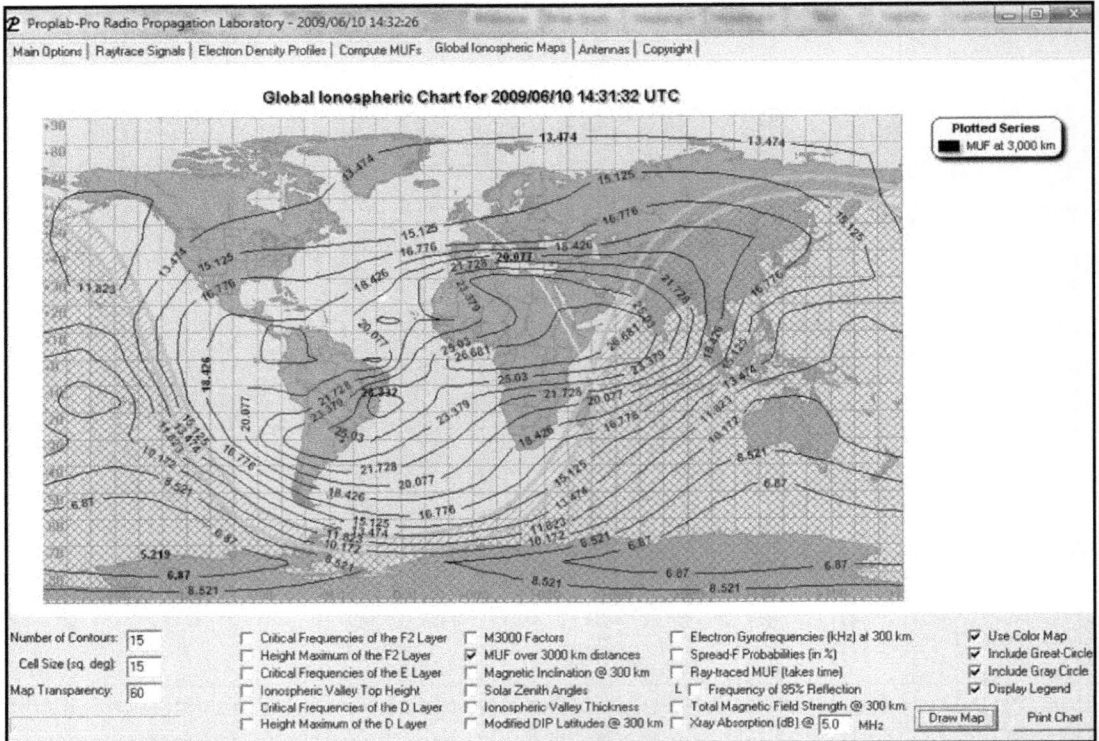

Fig. 5.21: World MUFs at 3000km.

wealth of information found on the NOAA sites, all about different aspects of propagation. Putting a few queries into Internet search engines produces good results.

The Australian Government's Bureau of Meteorology web site (www.ips.gov.au/) is another excellent place to start. It contains a wealth of information. At one time they offered a course on the subject of HF Ionospheric Propagation, but I could not find it whilst writing this.

Another site that offers a lot of useful information is Solar Terrestrial Despatch, (http://solar.spacew.com/). They also have a comprehensive course (very expensive) that will teach those with some previous knowledge a lot more about the subject.

The last but not least way I would suggest those wishing to gain more knowledge is to visit the local library, especially if it is within a large town or city. My own has a number of good reference type books.

Summary

This chapter has attempted to show the usefulness of computers in the shack as far as propagation is concerned. The authors of the various computer programs are to be congratulated for their hard work and for bringing such tools to a wider audience. Most of the programs mentioned can be enhanced upon, even if it is just organizing the output into something more individually meaningful such as averaging or producing medians. There is plenty of scope for the individual amateur to put their mark upon the whole.

6
Terrain Modelling for HF

by Alan Hydes, G3XSV

When talking about how good a site is for DX on the HF bands we sometimes hear statements like 'I have a good take-off to the west', or 'I get good reports from the West Coast of the US'. But is there a way of comparing sites or finding out how much gain or loss you are getting from the ground surrounding your antenna? The simple answer is yes... and the results can be quite astounding. This section will describe a practical method and data sources that can be used to analyse a site or compare sites and their DX potential. We will try to answer these questions:

- What angle of elevation is needed for a specific area of the world?

- How is the local terrain affecting the signal?

- What is the best height for the antenna, and can it ever be too high?

- How will moving the antenna change things?

The methods described here are based on software written by N6BV that has been published by the ARRL in their Antenna Book (24th edition). If you want the software, called HFTA (HF Terrain Assessment), it is contained on a download that accompanies the book (**Fig 6.1**). Here I will just give an overview of how to use it, how to get terrain data for a particular QTH, and show some example results.

The Basics

Most of us are familiar with antenna gain being shown as E-plane (or horizontal) radiation patterns and H-plane (or vertical) radiation patterns. The horizontal plane is often shown as if the antenna is in free space. For example, the radiation pattern of a dipole is shown in **Fig 6.2**.

In the H-plane (vertical), several radiation patterns are usually shown, because antenna height has great effect on radiation pattern. This is shown in **Fig 6.3**.

Modelling Terrain Effects

The HFTA software was developed to model how reflections and refractions from the local terrain (see **Fig 6.4**) add up to affect the resultant signal at a remote location.

Unfortunately the way that vertical an-

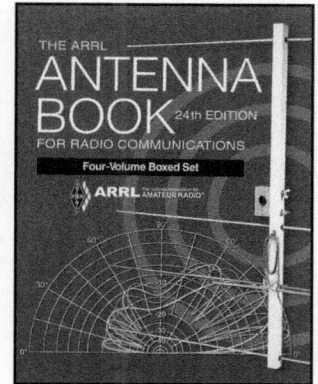

Fig 6.1: The ARRL Antenna Book.

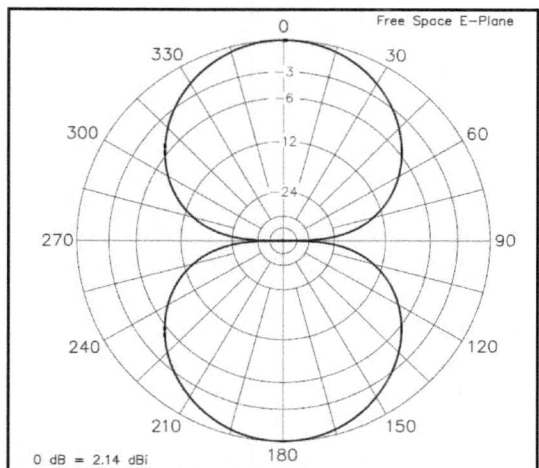

Fig 6.2: E-plane free-space radiation pattern of a dipole.

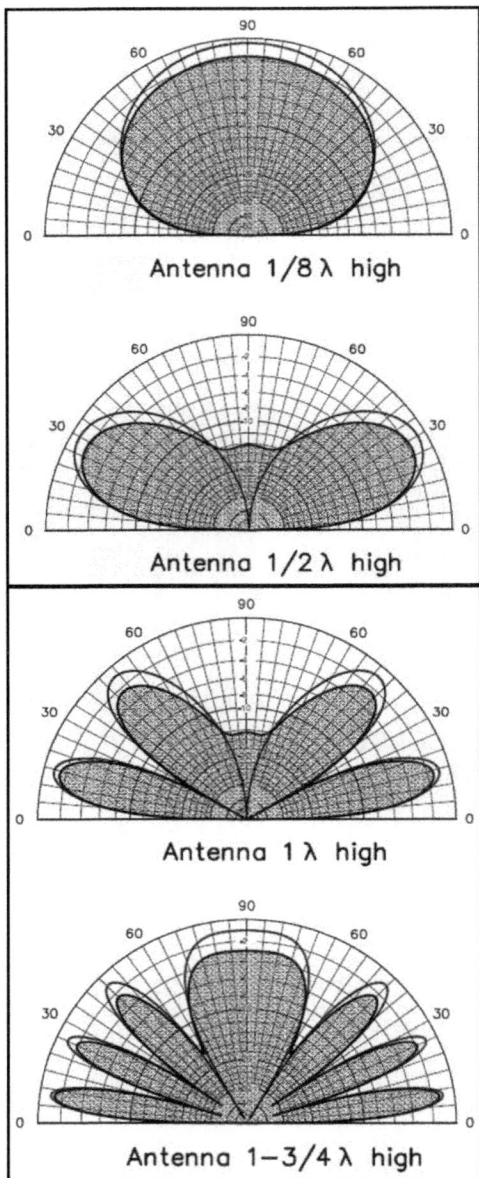

Antenna 1/8 λ high

Antenna 1/2 λ high

Antenna 1 λ high

Antenna 1-3/4 λ high

Fig 6.3: H-plane radiation pattern of a dipole at various heights above ground.

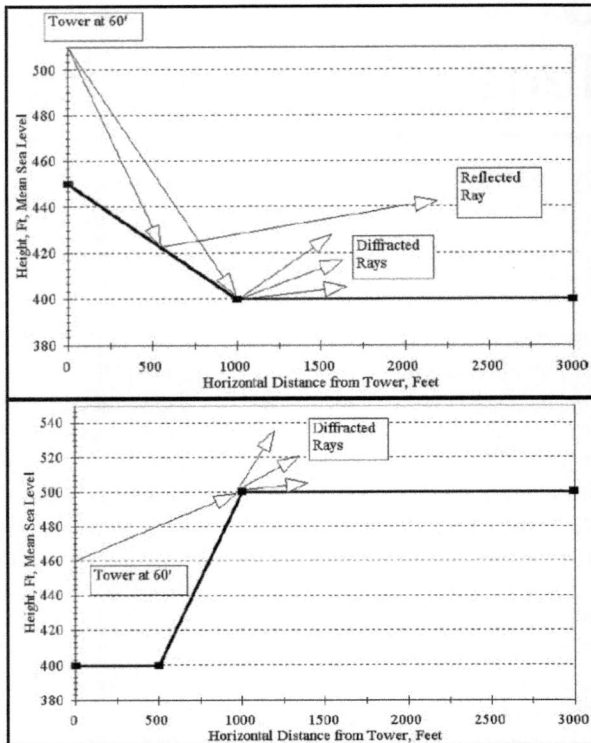

Fig 6.4: Diffracted and reflected rays.

need to provide it with the antenna height and the ground height for a particular direction over a distance of a few kilometres from the antenna.

Getting Terrain Data

The terrain data needed is simply the ground height above sea level along a line starting at the centre of the antenna and extending out several kilometres. HFTA needs a text file that contains two values on each line: distance and height. These can be in metres or feet. There are several sources of such data.

Printed Maps

The most readily accessible terrain data for the UK is Ordnance Survey maps. You will need to measure along a line in the direction you want to explore (for example 300° for the USA) For each data point you need to enter the distance and the height into a line of a text file and give the file an extension of

tennas interact with the surrounding terrain is much more complicated and the HFTA software does not attempt to deal with them.

Consequently the main function of HFTA is to calculate and plot the gain of yagi and dipole antennas for different take-off angles over a real ground profile. To do this you just

'.pro'. If you use metres as your unit of measure, you will need to put 'meters' (American spelling) as the first line of the file otherwise feet will be used.

Online Maps

An alternative is the use of an Internet accessible map which has built-in terrain data. An example is Google Earth. Unfortunately there does not seem to be a way of accessing the data programmatically, so the process for extracting the height is still quite lengthy. In Google Earth you can define the line along which you want to collect distance and height data by clicking the 'Add Path' function button. Then click the start point at your antenna location and an end point a few kilometres away in the direction you want to model. A white line appears on the map and you can now select 'Show Elevation Profile' from the Edit menu. This will show a profile similar to **Fig 6.5**. You can now move the cursor along the elevation profile and record the distance and elevation data point in a profile text file as for printed maps above. About 20 data points was found to be adequate, though 100 is better. Points can be closer spaced near to the antenna, to get better detail close in.

Digital Elevation Models (DEMs)

Another approach is to use digital elevation data that is available online. DEMs covering the UK are mostly provided by commercial enterprises and the data can be quite costly.

However, there is one DEM source that is public domain, as it was created by NASA. In February 2000 NASA ran the Shuttle Radar Topography Mission (SRTM) aboard STS-99. Raw data is in binary form and needs processing to produce the .pro files we need for HFTA. Hopefully someone will write a program to do the necessary transformation directly. Meanwhile, a method of using this data is described at the end of this section

What Take-off Angle?

Government communication agencies and major shortwave broadcast stations have been studying take-off angle for many years, in an effort to enhance the strength of their signal into specific parts of the world. Software called IONCAP was developed to simulate the action of the ionosphere at HF frequencies. But as IONCAP is too complex to use directly, calculations were done for all times and stages of the sunspot cycle. The results were aggregated into a small set of data files that are representative of paths between various countries and continents and it is these files that are used by the HFTA software as a reference.

Using the HFTA software

Fig 6.6 shows a screen shot of the main HFTA window, which allows you to enter the names of up to four files containing terrain data, plus the type and height of the antenna. In this example a terrain data file that represents flat ground has been used, in order to do comparisons with typical gain data seen in antenna textbooks. The other three terrain files

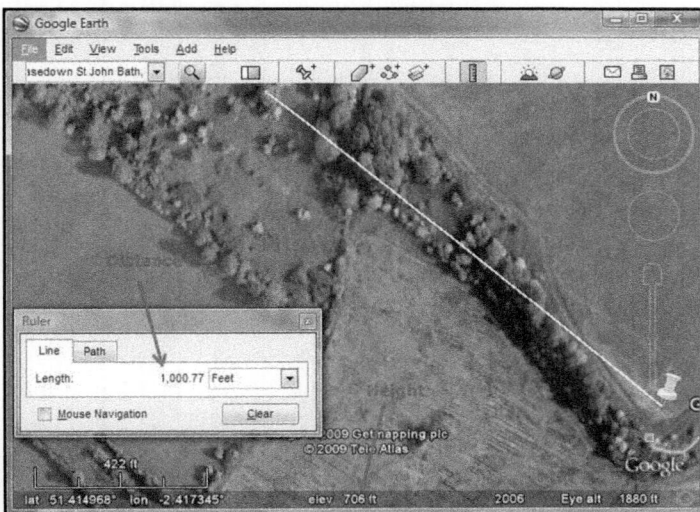

Fig 6.5: How Google Earth can be used to measure height at various distances from a given location.

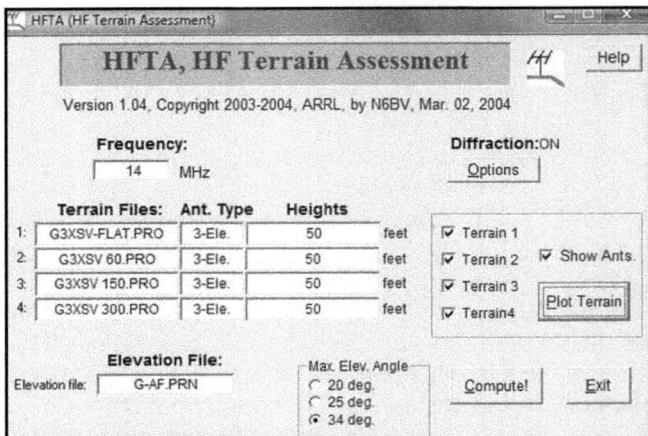

Fig 6.6: Main HFTA window.

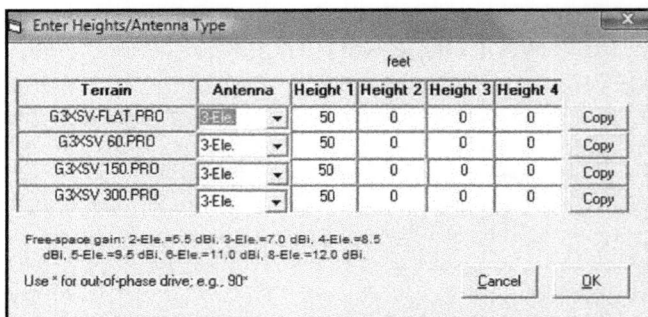

Fig 6.7: Antenna height and type dialog box

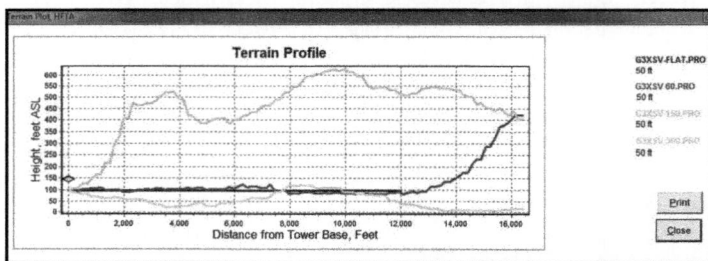

Fig 6.8: Plotted terrains at bearings of 60°, 150° and 300°, plus flat ground.

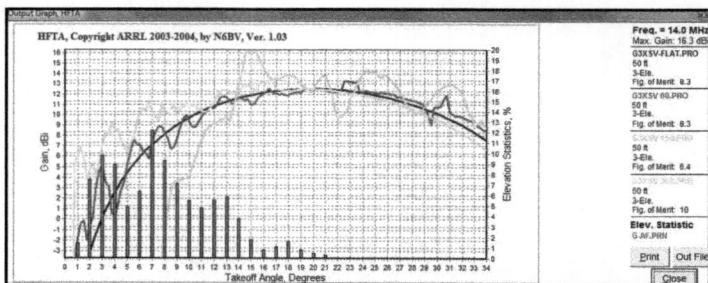

Fig 6.9: Plotted terrains at bearings of 60°, 150° and 300°, plus flat ground.

contain data for difference paths, representing directions 60°, 150° and 300°. 'Ant. Type' or 'Heights' can be changed by clicking on the field. This brings up a dialogue box shown in **Fig 6.7** to make your selection.

Note that more than one antenna height can be entered in a row. This models stacked beams. We've entered a single three element yagi with the antenna at 50ft.

Terrain Plots

If you select the tick boxes and click the 'Plot Terrain' button you will get a window showing a representation of the terrains shown in **Fig 6.8**.

In this example you can see that the antenna is above land that is approx 100ft and the elevation rises steeply along a path of 150° to a hill that peaks at over 600ft. This is in contrast with the path towards the USA at 300°, where the land steadily drops over the first mile.

Elevation Responses

On the main window you can select a particular part of the world, to see what take-off angles are likely to be useful for that path. In the example we have chosen a path from the UK to Africa which would be appropriate for the 150° path. Click on 'Compute'! The analysis results are shown in **Fig 6.9**.

In this example the most useful elevation angles for Africa are shown as a bar graph. You can probably spot how poor the QTH chosen is for propagation to Africa! Compared to flat ground, the gain is negative all the way up to 14° of elevation, which accounts for most of the propagation to that part of the world.

This is in contrast to the gain for the path of 300°, towards the USA, which shows enhanced gain compared to flat ground. This is between 6dB and 10dB at lower angles.

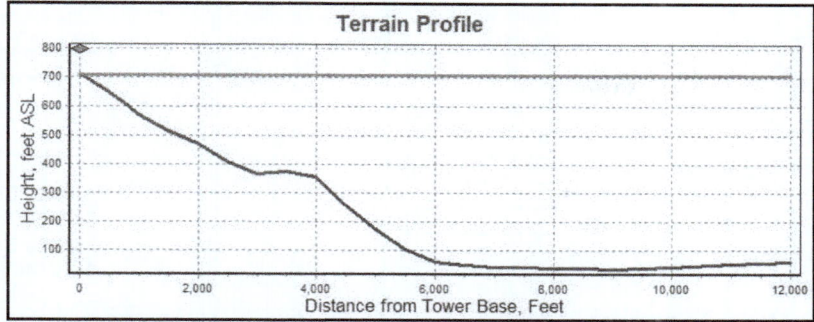

Fig 6.10: Terrain profile towards the USA from the Bristol CG's site (descending line) and flat ground (straight line).

Example and Interpreting Results

Here is an example to show what kind of gain is achievable from the best locations. This site in question is used for Field Day contests by the

Bristol Contest Group. It is located on the edge of an escarpment that slopes away to the west. **Fig 6.10** shows the terrain profile towards mid-west USA, showing the antenna at 90ft.

In **Fig 6.11** is the HFTA elevation response for this path, using a 3-element yagi at 90ft. Moving from left to right, the next line is for the same antenna at 90ft above flat ground and the third line is the same antenna at 50ft, to represent a typical home installation.

The bar graph shows the elevation statistics for take-off angles that are useful for the USA. It can be seen that for very low take-off angles of 1° or 2°, the gain relative to the same antenna at home is 20dB or more.

At angles under 6°, which are typical for west coast USA, the antenna over the slope significantly outperforms an identical one over flat ground. However, over 8° the performance drops off. This shows that the antenna may be too high for general coverage of the USA and a lower one would be an improvement for some of the

higher angle propagation. If it is practical, a better solution is to stack antennas and switch (or split) the power between them.

Conclusions

The combination of the analysis capability and the elevation statistics derived from IONCAP make the HFTA software and excellent tool. It can provide valuable evidence when comparing sites and when trying to determine what height and location would be ideal for antennas. As already mentioned, it can't deal with verticals, but it is nevertheless a valuable tool for any amateur who is serious about picking a site or optimising their antenna system.

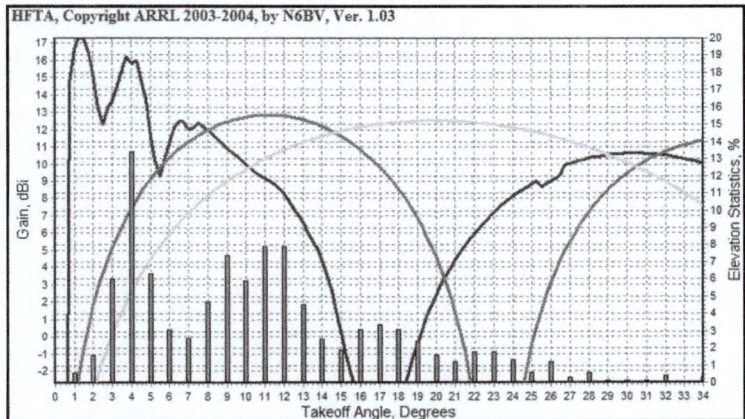

Fig 6.11: Gain of a 3-element Yagi at the Bristol CG's group's site, compared to an identical antenna at 50ft and 90ft above flat ground.

	A	B	C	D	E	F	G	H	I	J
1	Start Point		Degrees N	Minutes	Seconds		Degrees W	Minutes	Seconds	
2			51	56	50.89		3	6	42.74	
3										
4	Angle		135							
5										
6	Range (m)		5000	m						
7							Distance Between Points			
8	Number of Coordinates		100				50.5050505	m		
9							Distance Between Points			
10							0.00054616	degrees		
11										
12	Id	DEC_LAT	DEC_LONG			Distance from Point 1 (Feet)				
13	1	51.9474694	-3.111872222			0				
14	2	51.9470832	-3.111486027			165.699				
15	3	51.9466971	-3.111099832			331.398				
16	4	51.9463109	-3.110713637			497.097				
17	5	51.9459247	-3.110327441			662.7959				
18	6	51.9455385	-3.109941246			828.4949				
19	7	51.9451523	-3.109555051			994.1939				
20	8	51.9447661	-3.109168856			1159.893				

Fig 6.12: Lat Long Calculator Excel spreadsheet.

Generating .pro files from SRTM data

HFTA has an extensive Help file, which explains how to generate the terrain data from different sources. However, many of these sources are only available for the USA. As mentioned earlier, there is a way to use data from the SRTM. It involves use of web-based software at: http://www.gpsvisualizer.com/elevation

The site consumes input files containing longitude and latitude points and generates height/elevation data for those points. The method we use in our application is to generate a .csv file using an Excel spreadsheet. The same spreadsheet is then used to transform the output data into the format needed for HFTA. **Fig 6.12** shows a screen shot of the start of this spreadsheet with data entry cells shaded.

The Excel spreadsheet is available at: http://g6yb.org/hfta/lat-long-calculator-v2.xls Incidentally, thanks to Matt Jeffery, M0MAT, for help in developing this spreadsheet.

The longitude and latitude of the start point can be found from Google Earth, or from some GPS receivers, to the required accuracy of seconds to two decimal places. Angle is the direction of propagation you want to simulate and can be found from any great circle map of the world. Range and Number of Coordinates can be left as is. Bear in mind that the SRTM data has a resolution of 30 metres, so there is not much to be gained by having points closer than that.

After entering the start point and angle in Excel, save sheet 2 as a .csv file. Go to http://www.gpsvisualizer.com/elevation , click [Choose File] and browse to your .csv file. Select Output type of 'Plain text' and Units 'U.S.' then click [Convert & add elevation]. Cut the height values out of the text box produced from this site and paste into cell A1 in sheet3 of the spreedsheet. Then save sheet 4 as a .PRN file. Change the file extension to .PRO and you are ready to use it in HFTA.

7
Software Defined Radio

by Andrew Barron ZL3DW

No form of radio is more tightly integrated with computers than Software Defined Radio (SDR). This definition of software defined radio illustrates the point.

"Software-defined radio (SDR) is a radio communication system where components that have been typically implemented in hardware (mixers, filters, amplifiers, modulators, demodulators, detectors, etc.) are instead implemented by means of software on a personal computer or embedded system." Dillinger, M., Madani, K. and Alonistioti (2003), Software Defined Radio: Architectures, Systems, and Functions.

There are other definitions of SDR, but I believe that this one best describes what we mean by software defined radio. The ITU definition is so broad that it includes radios with any kind of computer connection, such as a CAT (CI-V) interface. These days, a superheterodyne radio with CAT frequency control is not considered to be an SDR. The FCC manages the radio spectrum in the USA. It is primarily interested in devices that emit signals on radio frequencies. Their definition says software defined radios must always include a transmitter. There is no FCC definition for software defined radio receivers. So, their definition excludes the many small-box, and SDR dongle, receivers that we do consider to be SDRs.

If we are to implement the various receiver and transmitter functions in software. We need a computer device of some sort to run the program. The computer that runs the SDR code does not have to be a PC (personal computer), it can be a microprocessor CPU (A computer processor used in many devices including personal computers) , an FPGA (Field Programmable Gate Array),An FPGA is a chip containing thousands of logic gates that can be arranged under the control of an external computer device to form very complex logic operations. There is also an ARM ARM. (Advanced RISC Machine. A reduced instruction set computer processor.) processor such as found in the Raspberry Pi and most other single board computers, a computer graphics card, or even the microprocessor in your cell phone.

The Digital Revolution

Like virtually all electronics technology, amateur radio receivers and transceivers have been caught up in the digital revolution. Your cell phone, TV, music recordings, and camera are digital, and so probably is a large part of your ham radio. What is the obsession with making everything digital? The most important factor is noise performance. All analog circuits in the receiver chain add noise to the signals that you want to hear or decode. But once a signal has been converted from an analog signal into a data stream of digital bits, it can be amplified, mixed, filtered, demodulated, or modulated without adding any additional noise.

The signal that your receiver picks up already has a combination of atmospheric noise and interference signals. The last thing we want is the receiver technology adding more noise and further degrading the signal to noise ratio. The goal for software defined radio receivers is to convert the received signals from analog to digital as close to the antenna connector as possible.

SDR transmitters convert from digital to analog as late as possible. Usually immediately before the final power amplifier. All of the audio, modulation, filtering, and frequency translation

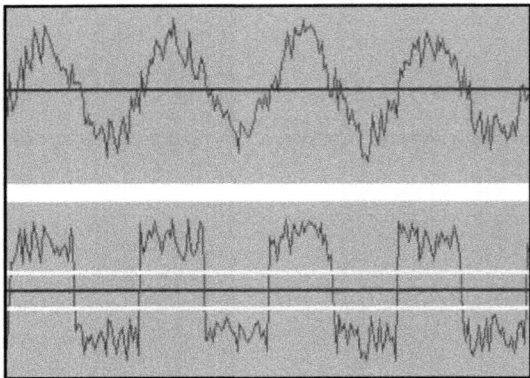

Fig 7.1: A noisy analog signal and a noisy digital signal.

takes place in the digital stages. It is diffiult to remove noise from an anolog signal. But a perfectly clean digital signal can be regenerated by sampling the digital signal at the threshold levels (white lines). If the level is higher than the upper level it will be regenerated as a binary 1 bit and if the level is below the lower level it will be regenerated as a binary O bit.

The second reason for 'going digital' is interference. Conventional hardware mixers and oscillators will mix any oscillator phase or amplitude noise onto your wanted signal and the non-linear mixing process introduces unwanted intermodulation and image signals. Software DSP (Digital Signal Processing). which is performed by manipulating numbers in data buffers, can perform the same functions as the hardware devices without introducing intermodulation distortion and unwanted signals. Phase noise or 'Jitter' on the ADC (stands for Analog to Digital Converter) and computer clock signal can cause problems, but the noise tends to have a fixed level and it will be well below the level of background noise that is received when the receiver antenna is connected.

The third reason is performance. Narrow hardware filters are difficult to make, and very careful circuit design is required to get good performance. Narrow filters are prone to ringing and they are usually fixed in bandwidth. Digital filters implemented in software can be variable and they don't exhibit ringing. Features like auto notch filters, variable high and low audio frequency filters, adaptive noise filters, I.F. width and I.F. shift controls, and band-scopes are common on radios, which use digital signal processing.

Another reason to 'go digital' is cost versus performance. It is very difficult and quite expensive to create a top-class conventional receiver. But the DSP part of a software defined radio is just a few computer chips, which can be mass-produced relatively cheaply. Firmware loaded into a DSP chip, or the FPGA easily updated, improved, and revised. Completely new modes can be added. For example, the latest transceivers can send and decode RTTY and PSK signals. With a hardware radio, the performance is fixed as a part of the initial design.

Your 1970s transceiver may have a terrific receiver, but it cannot be improved further without adding more hardware such as INRAD filters.

It is true that early AF DSP filters did introduce a variety of weird-sounding artefacts, but the technology is much more developed now. You may still be able to adjust DSP filters to the point where the wanted audio becomes distorted, but you don't have to.

Types of SDR

There are a myriad of different SDR types. USB dongles, small box QSD (Quadrature Sampling Detector. Used as a single conversion receiver or transmitter). based receivers and transceivers, SDR kits, VHF/UHF/SHF 'hybrid' SDR boards that use a mixer to get an I.F. frequency low enough for direct sampling, HF transceivers using direct digital sampling, SDRs with knobs and onboard DSP, Superheterodyne radios with IQ ('incident' and 'quadrature' signals treated as a single entity), output (at audio) from their DSP stages, and commercial or military radios that use digital sampling and DSP chips.

Some software. defined radios have dedicated software specifically written to match the hardware. Others can be used with a variety of different software applications. This is one of the advantages of SDR technology. You can completely change how your radio looks and operates simply by using a different software package.

Different software may offer different modes. For example, one program may include DAB or DRM digital broadcast reception, another might include FM, AM, and SSB, or a PSK decoder. Using an SDR that has front panel controls is much the same as using a receiver or transceiver based on the conventional superheterodyne architecture. Using an SDR connected to a PC is no more difficult than using any other computer program.

Dongle SDR Receivers

Dongle SDR receivers have a similar physical format to USB memory 'flash drives.' They are designed to be plugged directly into a USB port on your computer. You will need to load a free SDR program such as SDR# (SDR sharp), HDSDR, or SDR Console, and you may have to load a USB device driver called Zadig and/or a data format interface utility called EXTIO.dll. Most SDR dongle receivers use the R820T/T2 as the tuner chip, and the Realtek RTL2832U chip as the analog to digital converter, digital down converter, and USB port interface. They typically cover frequencies from 24 to 1766 MHz or up to 2200 MHz in some models. Some are bundled with cheap up-converters to provide coverage of the HF bands.

The maximum panadapter bandwidth is typically about 2.5 MHz. RTL dongles do have a filter in front of the RF preamplifier, but they don't have separate bandpass filters for individual bands. That makes them prone to overload and ADC gain compression.

The 8 bit ADC means that they don't have a particularly large dynamic range, to begin with. However, they are very sensitive, and they cover a huge frequency range. Most importantly they are very cheap. Because of the low price, they provide a great introduction to software defined radio and, they are extremely popular.

The 'FUNcube Dongle Project' came from a desire to bring radio and electronics to young people in an exciting way, using low cost hardware and software. As is often the case with this type of project, the resulting receiver has attract-

Fig 7.2: A typical dongle type SDR receiver.

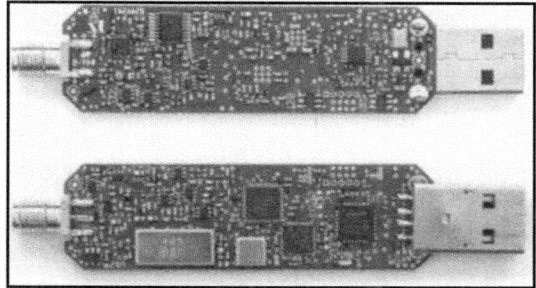

Fig 7.3: A look inside the FUNcube dongle SDR receiver.

ed great interest from enthusiasts and hobbyists all over the world. The FUNcube Pro+ dongle receiver uses a similar design, but a different chipset from the RTL dongles. It uses a Mirics tuner chip to down-convert the radio spectrum, followed by an SDR receiver. It also has better filters, including SAW (surface acoustic wave) filters for the 2m and 70cm amateur radio bands, a more stable TXCO reference clock, and a 32 bit ADC.

The radio has a frequency range of 150 kHz to 260 MHz and 410 MHz to 2.05 GHz. Covering every ham band from the 1800m band to the 23cm band. However, the FUNcube radio only supports a maximum panadapter bandwidth of 192 kHz, compared to about 2.5 MHz using an RTL dongle.

Seventeen amateur radio bands from L.F. to 'L band,' from a receiver you can carry in your pocket would have been unthinkable before the advent of SDRs.

The Mirics MSi001 tuner outputs I and Q audio streams which are converted to digital streams with an MSi2500 dual-channel 10-bit ADC, DSP and USB interface chip.

Small Box SDR Receivers and Transceivers

There is a big range of small box SDR receivers and a few small box transceivers on the market. Some use QSD technology but most are hybrid SDRs or use direct digital sampling. Some cover the LF, MF, and HF bands while others such as the SDRplay RSPdx shown in the picture, include coverage of the VHF and UHF bands as well. The RSP2 covers 10 kHz – 2 GHz with a 10 MHz maximum bandwidth. It can also be used as a spectrum analyser using free software.

HF small box receivers models usually use direct sampling with the DSP performed in an FPGA. The VHF/UHF models typically use a tuner chip such as the Mirics MSi001 RF tuner with the MSi2500 'digital interface' chip.

The direct sampling radios usually have a 12 bit, 14 bit, or 16 bit ADC. The ADC in the hybrid radios can be a much slower specification as they are working with lower input frequencies. They usually sample at either 16, 24, or 32 bits. There are some exceptions. The RTL dongles and HACK RF boards are only 8-bit radios. Many small box radios can support panadapter bandwidths of 8 MHz or more.

All dongle and small box radios rely on an external computer (usually a PC) to perform the control and DSP functions. They are often bundled with their own free SDR software, and many can also be used with SDR# (SDR sharp), HDSDR, SDR Console, or other SDR programs. The interface to the computer is usually a USB 2.0 or USB 3.0 connection, but some radios use an Ethernet connection or WiFi. Almost all SDR software is designed for Windows PCs, but there are options for macOS and Linux. Increasingly small box and PC board SDRs are being used with single board ARM-based Linux computers and Android phones.

SDR Tranceivers

There are far fewer SDR transceivers on the market. They are harder to design and manufacture and the market is smaller. Also, receivers benefit from the biggest performance gains.

The SunSDR2 pro from Expert Electronics is a 20 Watt HF, and 8 Watt VHF, SDR transceiver, covering a frequency range from 90 kHz to 66 MHz, or 95 MHz to 148 MHz. It is a direct sampling radio with 16 bit ADC sampling at 160 MHz. Interestingly, it uses the ADCs second Nyquist zone when it is operating in the VHF mode. With the free bundled 'SDRuno' software, the radio supports two 312 kHz wide panadapters and a lower resolution wideband display to 80 MHz. CW generation is performed in the radio rather than in the PC software, for low

Fig 7.5: SunSDR2pro transceiver.

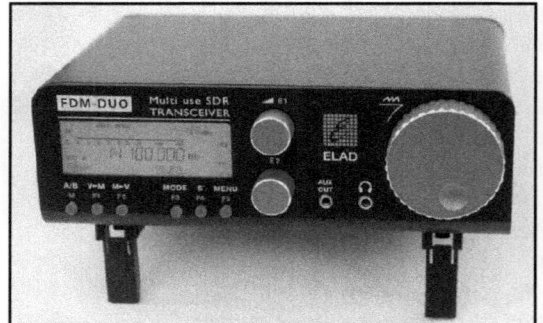

Fig 7.6: ELAD FDM Duo transceiver.

Fig 7.4: A 'small box' hybrid SDR receiver.

latency. The radio uses an Ethernet connection to the controlling PC.

The ELAD FDM Duo crosses the boundary between radios that need a PC for the controls and DSP to including the DSP and controls in the radio. It can also connect to SDR software running on a PC. It covers the HF and 6m bands and has a minimum transmitter output of 5 watts, The radio uses 16-bit direct digital sampling at 122.8 Msps (1 Msps is a million samples per second. 1 ksps is one thousand samples per second).

FlexRadio has been involved with SDR since the very start. They produced the first commercially available amateur radio SDR transceiver, the SDR-1000, in 2003. Rather than moving from the production of conventional superheterodyne architecture radios into the SDR world, FlexRadio has graduated from making black box SDRs that need a connection to a PC, to a range of superb 100 watt SDR transceivers that can be controlled with the front panel controls, via an external 'Maestro' control unit, with a PC running 'SmartSDR,' over the Internet, or on a phone or tablet computer. The FLEX-6600M 'Signature' range transceiver is a 100W SDR transceiver

Fig 7.7: FlexRadio FLEX-6600M.

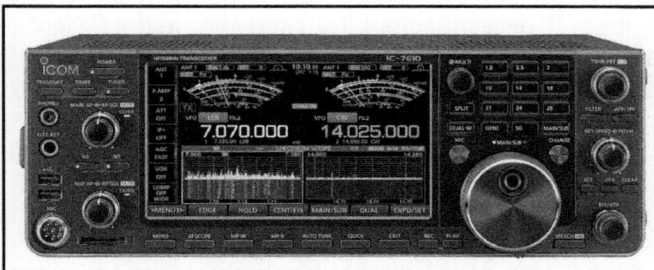

Fig 7.8: The Icom IC-7610 is a direct sampling SDR.

that can be used with or without a PC.

It can be used completely stand-alone, or you can use the 'SmartSDR' software. 'DogPark SDR' is a macOS alternative.

The FLEX-6600 version has identical performance but lacks the front panel controls. It is ideal for those who prefer one of the other methods of control.

The Icom IC-7610 is a 100 Watt HF +6m transceiver with two identical receivers and a large touchscreen. Each receiver can have a panadapter and waterfall display. It is a direct sampling SDR that does not require a connection to a PC, although you can connect the radio to a computer for CI-V control and digital mode operation.

QSD RADIOS

The early SDRs used quadrature sampling detectors (QSD), a technology pioneered and patented by Dan Tayloe N7VE in 2001. Although they work well, they are less common now, having been largely superseded by cheap hybrid designs. The Tayloe QSD detector is simple. It uses three low cost integrated circuits, a dual Flip Flop D type latch configured to divide the local oscillator signal by four, a multiplex switch, and a dual low noise Op-Amp.

More complex versions add a variable local oscillator clock to extend the frequency range. The most popular QSD boards are the SoftRock radios. The Flex-1500 and Elecraft KX3 transceivers are also QSD designs. Both have excellent receivers.

The QSD is an envelope detector. It acts as a single conversion receiver converting the incoming radio spectrum directly to audio frequencies. It is unusual because the I and Q streams are created as analog audio signals. They are converted to digital signals by the two ADCs on the stereo sound card in the PC.

The dual multiplex switch chip switches the incoming RF to each of the four outputs on every cycle of the

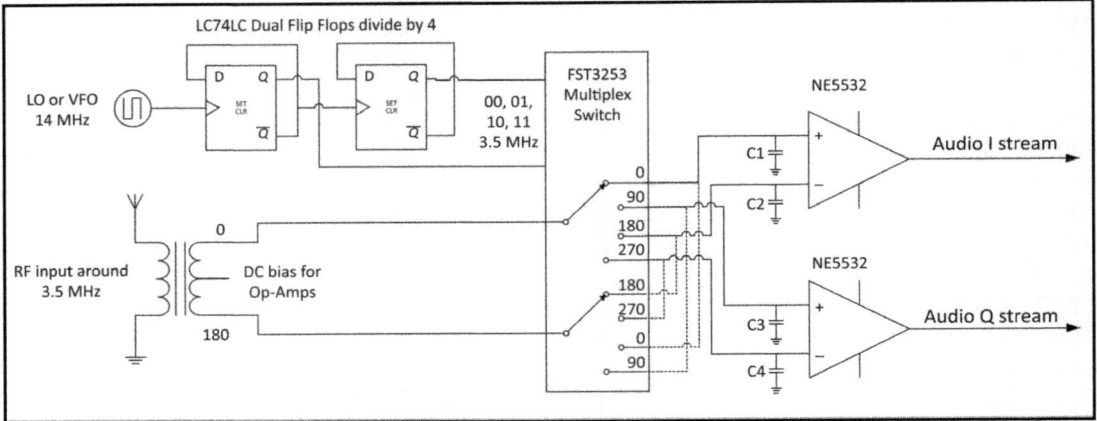

Fig 7.9: Schematic of the double balanced Tayloe quadrature sampling detector.

incoming frequency. In position one, C1 charges up according to the received signal voltage. C2 charges up with the inverse voltage because it is connected to the other side of the input transformer. The op-amp adds both signals together. The capacitors 'store and hold' the voltage while the switch moves through the next three positions. In position two, C3 charges up according to the received signal voltage and C4 charges up with the inverse voltage. Because this happens a quarter of a cycle later, the voltage being stored and summed by the Op-amp is 90 degrees later than the voltage on the first op-amp. When the switch reaches the third position, C2 and C1 are topped up with the level of the incoming signal. By now the input signal is at the 180 degrees point, so the connection to the capacitors is reversed. 90 degrees later when the switch is in position four, C4 and C3 are topped up again. The capacitors store the charge much like the capacitor following the diode in a crystal set. On the next cycle of the incoming RF, the process repeats. The output of the detector is two audio signals with a 90-degree delay on the 'audio Q stream'. The in-phase signal is called the incident stream (I) and the 90 degrees signal is called the quadrature (Q) stream because it is at 90 degrees when shown on a vector diagram.

The block diagram is shown in drawing E in **Fig 7.10**. If you want to transmit as well as receive, a QSD can be used in reverse as a QSE (quadrature sampling exciter).

Direct conversion receivers are like the Superheterodyne receivers we are familiar with, except the I.F. output is directly at audio frequencies extending from 0 Hz up to the bandwidth of the receiver.

All direct conversion receivers suffer from image signals. If we say that the nominal receiver frequency is FO = 3.500 MHz as shown on the schematic. A signal received at 3.501 MHz will result in a 1 kHz audio output from the detector. A signal received at 3.505 MHz will result in a 5 kHz audio output from the detector. That is great! The problem arises when a signal below the nominal receiver frequency is received. For example, a signal at 3.499 MHz. The detector cannot output a minus frequency, so it folds the signal over and outputs it at plus 1 kHz. This is a very big problem because the signals received at frequencies below the nominal receiver frequency will interfere with the ones received above it. To make things even worse, the sidebands will be reversed so an LSB signal will come out as USB, and the lower the received frequency, the higher it will appear on the detector output. Luckily, we have the Q signal. If we apply another 90-degree phase change to either the I or the Q signal and sum the I and Q streams, the wanted signals above FO will be enhanced and the unwanted image signals will be cancelled. If the two streams are subtracted, we can extract the signals from below FO and suppress the signals from above FO. That way we can show the signals above and below the centre frequency without any image signals.

I and Q signals are used in all SDR receivers to allow the suppression of image signals. The difference between the two streams is also used to demodulate (or modulate) phase (or frequency) modulated signals such as FM. You can demodulate amplitude modulated signals such as AM, CW, or SSB from either the I or the Q stream.

DSP Evolution

Software defined radio is a logical extension of the digital signal processing (DSP) found in many amateur radio transceivers and some shortwave receivers. Digital signal processing in conventional receivers and transceivers is performed by specialised DSP chips in the radio. They are programmable devices running software ('firmware') like a microprocessor, but they are specially designed to manage radio, audio, or video, signals.

AF (audio frequency) DSP works by converting the audio output from the demodulator into a digital signal. The digital signal is manipulated to apply filtering and noise blanking and then it is converted back into an audio signal and passed on to the audio amplifier stage. With I.F. (intermediate frequency) DSP, the output of the first, second, or third mixer of the receiver is converted into a digital signal. The operating frequency may be in the low RF or high audio frequency range, but in this case, the demodulation is carried out as a digital process in the DSP stage of the receiver. In a transceiver, the modulation would also be performed in a DSP stage. Features like I.F. shift and I.F. width control are performed using firmware code running on the DSP chips and there will usually be noise and notch filters. Often a band scope is provided so that you can see a display of signals above and below the receiver's frequency. A software defined radio receiver takes the DSP process even further by performing the analog to digital conversion after the receiver preamplifier, or after a single down-conversion mixer. Software defined radio transmitters use digital signal processing to modulate and shape the radio signal.

Software defined radio is a merging of radio technology and computer technology. Some of the radio is still hardware circuits but the operating panel and DSP are managed by software applications running on computer devices in the radio, or external to the radio in the case of an SDR coupled with a personal computer.

A direct sampling SDR has no analog mixer stages. The RF signal is sampled directly at the incoming RF frequencies. A hybrid SDR usually has one down-converting superheterodyne mixer and oscillator before the ADC.

The ADC

The analog to digital converter (ADC) is a critical component of any direct sampling SDR receiver. But its performance is much less important in hybrid SDRs because the sampling is happening at a lower frequency.

Every time the clock signal goes high (or sometimes low), the ADC measures, or 'samples,' the level of the radio spectrum at the input, and outputs a number that relates to the input level at the sample time. It does it very fast. The LTC-2208 found in many direct sampling SDRs is clocked for 128.88 million samples per second. For each sample, it outputs a 16-bit number as a parallel data stream on 16 output lines. Low-speed ADCs output a single serial data stream, but this is impractical for an ADC sampling at 122.88 Msps because the data rate would have to be 1966 million bits per second. Some 16-bit ADCs have 8 output lines. They output the 8 bits of information when the clock signal goes from low to high and the remaining 8 bits when the clock signal goes from high to low. This does require some clever data buffering with a shift register, but it reduces the number of pins on the chip. Most of the very fast ADC chips used for direct sampling SDRs output 14-bit or 16-bit data.

ADC chips in hybrid receivers only have to sample at a minimum of twice the highest frequency in the I.F. passband, so they can use much cheaper and slower ADC chips. Their sampling rates range from the 96 ksps that a PC soundcard can perform, up to around 40

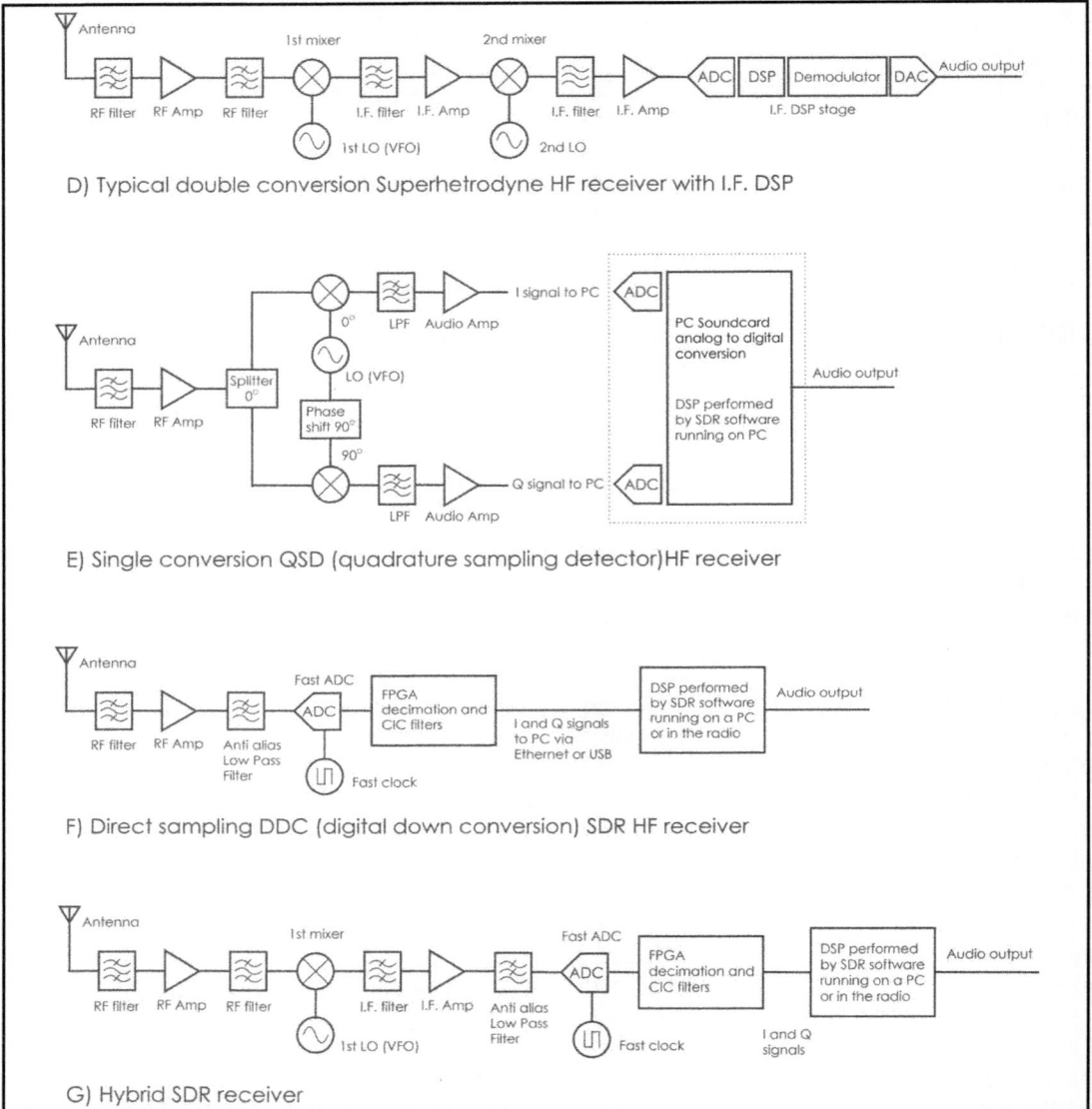

D) Typical double conversion Superhetrodyne HF receiver with I.F. DSP

E) Single conversion QSD (quadrature sampling detector)HF receiver

F) Direct sampling DDC (digital down conversion) SDR HF receiver

G) Hybrid SDR receiver

Fig 7.10: The evolution of DSP receivers.

Msps. In some cases, the I.F from the mixer stage is split into I and Q analog signals at audio or I.F. frequencies, and two ADC chips or a stereo CODEC (coder/decoder) chip are used for the analog to digital conversion. The I.F. frequency applied to the ADC in a hybrid SDR can be quite low. Some radios sample at 24 kHz. The Yaesu FTDX101 and FTDX10 transceivers sample the 9 MHz I.F.

Harry Nyquist (1889-1976) determined the fundamental rules for analog to digital conversion. He found that as long as the sample rate (FS) was at least twice the rate of the highest frequency in the analog signal, the original analog signal could be recreated accurately from the digital data. This means that to accurately, recreate a radio signal at 50 MHz (FS/2), we would need to sample at a minimum of 100 Msps (FS). The range of sampled frequencies from 0 Hz up to FS/2, is called the first Nyquist zone. With suitable filters, the ADC can safely sample frequencies

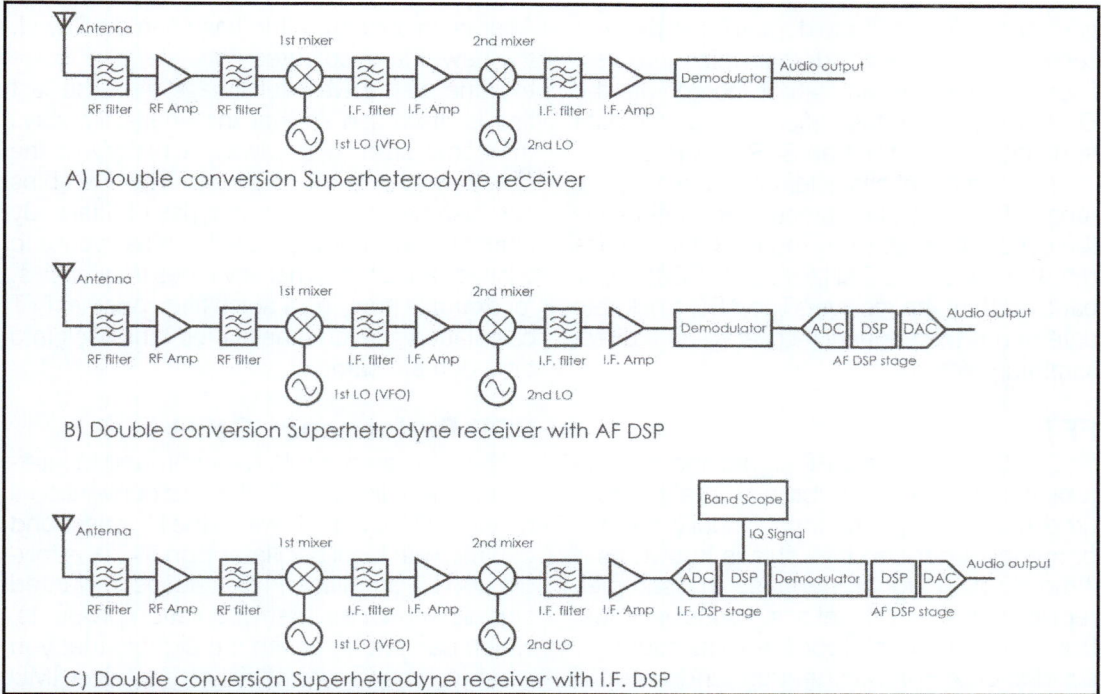

Fig 7.11: The evolution of SDR receivers.

above FS/2. The idea of deliberately receiving frequencies above the first Nyquist zone is called 'undersampling.' It is used in some receiver designs. For example, the SunSDR2 pro radio has an ADC sample rate of 160 Msps. It uses its 2nd Nyquist zone to provide coverage of 80-160 MHz. The radio switches automatically between HF and VHF modes when you tune to a frequency above 80 MHz. However, it is more common to use a hybrid SDR design than an undersampling design because high-speed ADCs are far more expensive than the ones used in hybrid receivers, and hybrids can cover a much wider range of frequencies.

You can only use one Nyquist zone at a time, i.e. HF or VHF, not HF and VHF. Makers must provide a low pass or bandpass filter before the ADC to prevent signals from other Nyquist zones from being sampled. If you try to cover more than one zone,

the signals from both zones will be sampled and they will overlap on the panadapter. Once signals from two or more Nyquist zones have been sampled, they cannot be separated again. Without a bandpass filter on the SunSDR2, a signal at 10 MHz would appear right on top of a signal sampled at 150 MHz.

Technically the whole Nyquist zone can be displayed on the panadapter (band scope) display at once, but that involves dealing with very high-speed data and a very fast FFT (Fast Fourier Transform. A complex software pro-

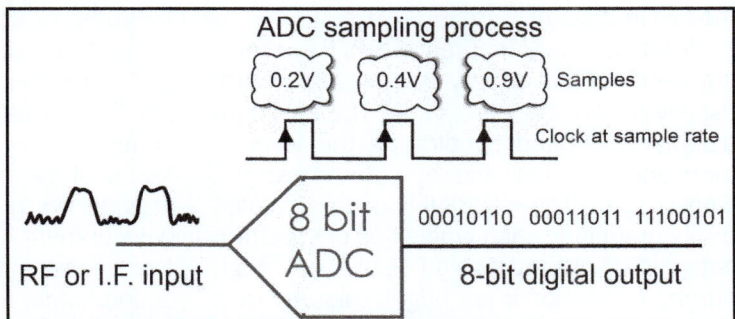

Fig 7.12: An example of ADC sampling.

gram that converts 'time domain' samples per second, into 'frequency domain' bins) process. It can be done in radios that incorporate the DSP process, but it is difficult to transfer data at nearly 2 Gbits from an SDR to a PC.

The number of bits determines the dynamic range of the sampling process and ultimately the dynamic range (maximum signal to noise ratio) in the 2.4 to 3kHz SSB, or 500Hz CW bandwidth of the receiver. The ADC clock rate determines the frequency coverage of a direct sampling receiver.

FFT

The ADC samples the RF signal and outputs a digital stream of numbers at a rate measured in samples per second. Usually, millions of samples per second. This is known as a 'time domain signal' because the samples represent the input level at successive times. But that is not useful for the panadapter or band scope. For the spectrum and waterfall display, we want to see a representation of the signals across a range of frequencies, known as a 'span.' A 'frequency domain' display.

The Fast Fourier Transform (FFT) is an exceedingly complex mathematical algorithm that transforms 'samples' in the 'time domain,' to 'bins' in the 'frequency domain.' The software is so complicated that most developers use a reusable software code block called a .dll (dynamic link library) written by someone else to do the FFT. Open Source SDR programs often use FTTW (Fastest Fourier Transform in the West). SDR radios with DSP onboard usually use a dedicated DSP chip which includes the FFT transform.

A 'bin' is a level that represents the incoming signal within a very narrow bandwidth, usually only a few Hertz wide. The spectrum display is created by plotting the level of thousands of bins side by side across the display span. The waterfall is usually created by accumulating each whole line of spectrum data and displaying it as a line of pixels. High numbers are plotted as bright coloured dots and low levels are plotted as darker colours.

After each spectrum line has been displayed, the new line of coloured dots is added to the top line of the waterfall image file. The last line is discarded so that the image file stays the same size. In this way, a history of the spectrum display is depicted. The FFT bins can also be used for some types of filters. By changing the numbers in the bins, you can easily change the signal at various frequencies, for example applying a notch filter. Another FFT calculation converts the filtered bins back into time domain samples.

SDR Performance Gains

SDR receivers are relatively immune to intermodulation distortion (IMD). In a conventional receiver, large signals within the I.F. passband can mix with local oscillator signals in the mixers and I.F. amplifiers. This causes unwanted signals known as intermodulation products, which can interfere with the signals that you want to hear. Susceptibility to receiver IMD is one of the main differentiators between an average receiver and a 'good' receiver. If you can't pick out individual stations working a busy contest and everything sounds mushy and distorted, don't blame the contesters. Poor receiver IMD is the most likely reason. It is easy to measure in conventional Superheterodyne receivers using the well-established 'two tone' test method. Software defined radios generally don't have mixers (in hardware) so the causes and effects of intermodulation distortion are completely different. In an SDR receiver, the IMD performance is not predictable and, in many cases, large signals near the operating frequency can actually improve the receiver's performance.

Software defined radios don't necessarily perform better than conventional radios and just like conventional architecture radios, you generally pay more for radios with higher levels of performance. However, SDRs do provide many new features and continual development. The recent transceivers from Icom, Yaesu and Elecraft indicate that software defined radio is here to stay.

8
Data Modes Software

by Sean Gilbert, G4UCJ

Introduction

Data vs. Digital

Before continuing any further, the matter of what to call these various transmission modes should be addressed to alleviate further confusion. In many publications and indeed software, the term 'digital modes', or 'digimodes' is used to describe those modes that are actually soundcard generated data, transmitted via an analogue audio system such as single sideband, AM or FM. These are actually data modes, rather than 'proper' digital (where the human voice is digitized via special codecs and converted to a bit stream before transmission). Digital voice systems such as those used on V/UHF repeaters, etc. i.e. D-Star, System Fusion (a.k.a. 'Fusion') and DMR are a separate topic altogether. In this chapter we will be dealing with sound-card generated data modes. At the end of this chapter there is a list of websites about some of the software that is mentioned.

History of Data Modes

Over the past 10 or so years since the last edition of this book was published, it is fair to say that the use of data modes in amateur radio has increased enormously. Historically, radio amateurs using data, rather than AM voice or Morse code to communicate came about as ex-commercial surplus mechanical TTY (TeleTYpe) machines, such as the Model 26, became available after the end of WW2. Machine generated RTTY (Radio TeleTYpe) remained popular amongst enterprising amateur community, right up to the point where computers became cheap enough to take over. Some nostalgic operators still use mechanical RTTY machines, such as the venerable Creed 7 series, but they require a good mechanical knowledge to operate and service. The machines are also rather noisy. The vast majority of RTTY operators now use computers to generate the required mark and space tones. When soundcards were added to the basic PC, they became multimedia devices capable of playing music, video and gave games a whole new life.

The simple, single note sounds used in games up to that point were replaced with complex multi-tonal sounds. Soundcards with audio output sockets meant better quality speakers could be used, replacing the tiny case mounted speakers found in those early computers. The audio-out and line-in sockets provided a pathway for inquisitive amateurs, who soon discovered that multi-tone data could be encoded and decoded by using the processing power of the new soundcards.

In 1998, Peter Martinez, G3PLX, building on the original system from Pawel Jalocha (SP9VRC), was one of the first to design a data mode that could be generated by computer and a simple soundcard, fed into the transmitter as an audio signal and then sent over the air using SSB (Single SideBand) in the same manner as a voice transmission. At the receive end, the audio output from the radio was routed into the computer soundcard via standard audio leads, and decoded by the specially written software. For better reception quality a simple interface consisting of an impedance matching network could be used, but were not essential. The more complex interfaces employed additional signal isolating transformers, transmit/receive switching and noise filtering components.

This new mode was designated as BPSK31 (Binary Phase Shift Keying) and has a transmission speed of 31.25 symbols per second. Rather than use the exact symbol speed, this was rounded down (or up) to the nearest whole unit to make the mode names more 'user friendly'. The transmission speed of these data mode signals is measured in 'bauds', or 'bd' (from the 'Baudot' code which was developed for the first digital systems, by Émile Baudot back in the 1870s), one baud being equal to a transmission speed of one symbol per second. PSK31, has a transmission speed of 31.25 symbols per second (31.25bd). PSK63 has a transmission speed of 62.5bd (i.e. double that of PSK31) and PSK125 is four times faster.

BPSK has a sibling, QPSK (Quadrature Phase Shift Keying). The main difference between the two is that BPSK is sideband independent, meaning it can be transmitted in either USB or LSB and the receiver can be in either mode without affecting the data decode. QPSK is sideband dependent, so both transmitter and receiver must be on the same sideband, else the signal will not decode. The advantage of QPSK is that it is more resilient to errors caused by interference than its binary relative. For the sake of clarity, we will refer to these collectively as 'PSK'.

The advantages of data modes over voice are that to recover enough data to complete a QSO requires a much lower SNR (Signal to Noise Ratio) than an SSB voice signal. The transmission speed of a mode dictates the SNR required for correct decoding to take place. The slower the data rate, the lower the SNR will be. When comparing data modes to voice, there is debate over just how many dB (deciBel) the improvement is. Having looked at various sources, there does not seem to be a definite answer, so the levels mentioned here are a rough average. If we take an SSB voice signal as the base line for communication, a PSK31 signal requires an SNR of around 10dB below the level required for a voice QSO. This 10dB difference (10x less signal reaching the receiver/decoder) between the two modes means PSK31 requires less power to conduct a successful QSO, or a QSO can be made when the band is not open enough to support SSB voice. The second advantage of data modes in general, over SSB voice, is that a single voice QSO requires about 2.4 kHz of the RF spectrum, whereas a PSK31 signal needs a little over 31Hz of bandwidth, meaning a standard voice 'channel' could, in theory, contain as many as 75 different non-interfering PSK31 signals.

This increase in efficiency over voice helped PSK31 become enormously popular. The mode was further developed to include faster variants such as BPSK63 and BPSK125, being twice and four times faster than BPSK31, but at the cost of using two and four times the bandwidth, respectively, of the original mode. 31.25. The transmission

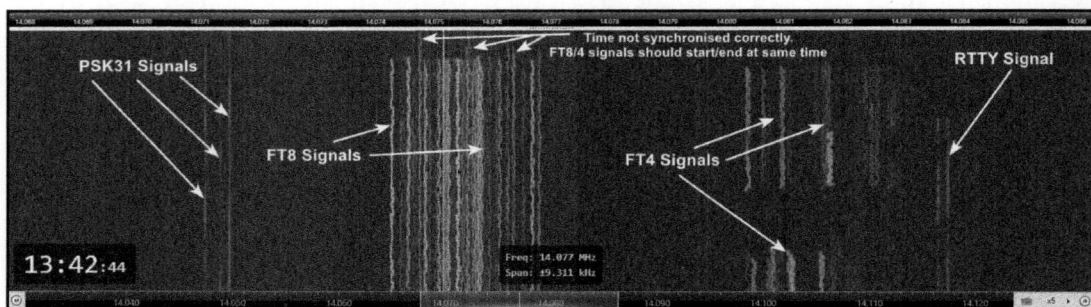

Fig 8.1: Example of PSK31, FT8, FT4 and RTTY modes.

rate of PSK31 was chosen as it equates to a typing speed of about 50wpm (words per minute), which is not too difficult to achieve, given some practice. To make things easier, whole blocks of text could be stored in files which could be recalled and inserted into the QSO as required, thus eliminating a good deal of the typing. These text files are called 'macro' files and are explained later in this piece.

In the following years, the floodgates opened and a multitude of different data mode types were invented and tested. Some became popular, whilst others never gained traction for whatever reason and disappeared almost as quickly as they appeared. To this day, modes are being developed, perfected and the cycle will continue for as long as computers are used in the transmission of data signals.

A Word of Caution

In this chapter, a selection of the most popular data mode programmes will be introduced, with installation instructions and a brief look at what each has to offer. A comparison table is included to help navigate through the modes offered by the various decoders. Some packages offer just a single mode, others may have a plethora of modes, including some that will be rarely selected but are included for specialist experiments. It should be remembered that it may not be legal to listen to, or attempt to decode, any signals you may hear that are either outside of the amateur radio allocations or transmitted by anyone other than a licenced radio amateur or licenced broadcast radio station. It is recommended the user consults the relevant authorities in their country to ascertain the legality of monitoring certain radio transmissions.

Is Downloading Safe?

Whilst there is always an element of risk when downloading software from the internet, there are a few simple steps that can minimize those potential hazards. Ensure the computer has solid internet security and/or antivirus/antimalware software that scores highly in relevant tests. Whilst some of the free ones are very good, you always get more comprehensive cover with the paid-for versions. For my own system, I use the built-in security and antivirus features of Windows 10, plus a couple of other programmes, both of which offer free and paid-for options. These solutions are often recommended in both computer publications and computer information websites. This combination has been sufficient to ensure that my PC has remained virus and malware free.

Besides having a solid antivirus protection, it's worth making sure you download from a safe site. In most cases that means downloading from the software author's own site or one that has been recommended by the author. Some sites that are promoting the software may not be totally safe. The links provided in this chapter are either the author's original or an approved distributor. The links have been personally checked and the downloaded files scanned to ensure they are virus free. For personal peace-of-mind, it is wise to scan everything you download, even images – just in case there is an additional payload secreted within. This is particularly true of so-called 'patched' or hacked versions claiming to offer paid software for free – they often have an unwanted malware surprise hidden within.

Installing Software

When installing software, be it radio related or not, each will have a default installation path set by the author. This path will, usually, have a destination within the 'Program Files' folder, for 64 bit software or 'Program Files (x86)' folder, for 32 bit software installations. Most 32 bit software will run on a 64 bit system but the reverse is not true. To run 64 bit software, the computer must have a 64 bit operating system installed. If you install to these program file folders, the UAC (User Account Control) will be triggered and, dependent on settings, may ask if you accept the installation of this software. The UAC stops the automatic installation of un-authorized apps and helps prevent software (or users) from making changes that could affect the system in a negative manner.

Some users choose to set the UAC to the lowest setting, which suppresses those warning messages and allows software to be installed without interruption. This does pose a possible security risk and it is not recommended. The software may be installed to a different folder, away from the 'Program Files' folders, if desired. It is left to the choice of the user where to install software, however, the default path set by the author is recommended to minimise any potential installation issues.

The majority of amateur radio software is designed to run under a Windows operating system, usually Windows 7, or later. Some programmes are compatible with earlier editions, such as XP or Vista, but these are now few and far between. One easily solved problem with Windows, is caused by the 'Smart Screen' filter that's built into some versions of Windows. This checks the software publisher and a few other things against a centrally stored list in an attempt to protect you from malicious software. However, most amateur software developers will never get on the Microsoft list, so the software gets blocked. If you want to run the software, you should click on the "More info" text link and then choose "Run anyway". It is essential that all software is scanned with, at least, antivirus software before installation.

There are programmes available for other operating systems, such as Mac and Linux. The use of simple, single board computers such as the Raspberry Pi (RPi) and Arduino has reached such a point that versions of some popular software have been written to run on these small, but surprisingly powerful computers.

Virtual Computing
If software is installed and uninstalled on a regular basis, the registry and other settings on a Windows PC will gradually become littered with remnants of past programmes, leading to the computer becoming less responsive and the operating system may become less reliable. If you want to try lots of software without compromising your main PC, virtual computing is the way to go. A virtual PC is a specialist software package that emulates a computer within your PC. When you run this application you are able to load Windows, Linux or any other operating system without affecting your main PC. Once the new operating system in installed you can connect to the Internet and download/install all the software you like without affecting your main PC. The software is actually installed onto a file that's owned by the virtual PC. Connections to the outside world for Internet, sound, printing, USB ports, etc. are all handled automatically by the virtual PC software, as it creates a bridge between the virtual PC and your hardware. The two most popular Virtual PC packages are the Oracle 'VirtualBox' & the VMware 'VMware Player'. Both are free downloads and work extremely well.

Data Modes Software Types
When it comes to data modes software, there are two main groups of software. The first group are the general purpose, multi-mode programmes that attempt to cover as many modes as possible within a single application. These are a popular place to start, as it is only necessary to familiarize yourself with a single set of controls to enable the operation of many different modes.

The second group are specialist programmes that only cover a single mode or a very small group of similar modes. These are often the only way to go for some modes (such as ROS), but do generally offer better results for their specific modes. You will often find that these packages have been written by the inventor/developer of the data mode in question, and for that fact alone they are likely to be particularly effective.

Equipment Required

Virtually all recent computers, regardless of size or complexity, have an on-board sound device. This can be used for data mode reception and generation but a separate plug-in card or external USB sound device is recommended for best SNR. The computer will need to be capable of running the relevant software, and have enough audio/com ports to allow connection to your radio. Most computers that have been produced in the past 5 to 10 years will run most, if not all, of the currently available software.

Older and lower performance computers and tablets may not be able to run certain software without a reduction of performance settings within the software. A good indication that the computer is falling short of performance is a general slow-down in software reaction times and often a 'stuttering' on the audio output. Should this happen, it may be necessary to reduce the number of programmes or browser tabs open at any one time. Software that uses FFT (Fast Fourier Transform) 'waterfall' displays can be particularly CPU intensive, especially at higher definition settings. A separate graphics card can, when using appropriate software, divert a good deal of the processing load required to generate the FFT display away from the CPU.

When using an SDR, you will need at least one VAC (Virtual Audio Cable). VAC's are a software emulation of a physical audio cable that routes audio from the output of the SDR software to the input of the decoding software. Another very useful addition to your system are VCP's (Virtual Com(munication) Ports) . A virtual com port is a software emulation of a physical serial com port. The addition of a number of virtual com ports allows the system to be configured such that each programme can be allocated its own port(s). Virtual com ports are configured in pairs, one port is used at the SDR software end and the matching port is used in the decoding software for such things as rig control, PTT switching, mode selection, etc.

To receive data mode signals, you need a method of receiving the signals. This can be a conventional receiver or SDR with your antenna system, or you could use an online web SDR and route the received audio to the software by a VAC. When using a conventional (non-SDR) receiver, the audio output of the receiver connects to the 'Line in'/'Mic in' of the computer soundcard.

Using the 'line out'/'rec out' or similarly marked socket on the receiver will ensure a fixed level signal, independent of volume setting, is fed to the soundcard.

If the PC has just a combined headphone/ microphone socket, or only a 'mic in' socket, the audio output level from the radio may be too high and could overload the decoder input. To avoid this, an in-line attenuator may be needed as line input/output levels are considerably higher than mic input levels (also there will be a considerable impedance mismatch, in the order of 4:1).

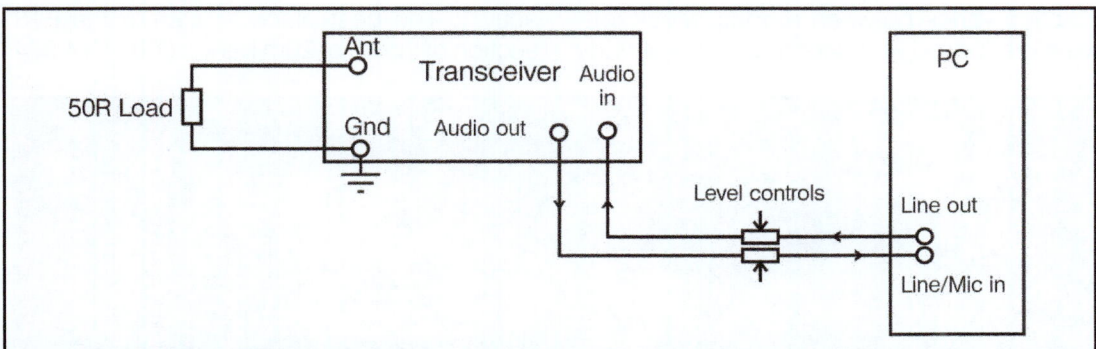

Fig 8.2: Connecting PC and radio.

What is a Macro?

Macro commands are little more than collections of stored text messages that may be recalled with a single key press. The use of macros are commonplace across most data modes, but newcomers are often unsure where to start. The use of macros can be intimidating at first, especially when the distant station bombards you with stacks of stored text and you have none ready!

The use of macros has made fast data modes very popular. If programmed correctly, it is possible to complete a 'rubber stamp' style QSO by entering the other station's callsign and name, then pressing various macro keys. The appropriate information is pre-filled by the software, using commands previously set in the macro.

The decoding software will have instructions on how to write macros for various tasks and modes. Some offer advanced features that can send commands to the radio and/or logging software so the QSO can be logged and all relevant information is transferred to the logbook.

Fldigi (and many others), have the facility where adding the characters '<TX>' automatically switches the programme to transmit and '<RX>' switches it back to receive. Some software allows you to auto insert details such as the 'QSO number', which the software obtains from the log. Many have methods of capturing the other station's call, which can then be automatically inserted into a macro. The actual macros required will depend on the mode in use, as operating practice varies between modes. Most software is supplied with generic macros already setup, so the user just needs to modify them to suit their operating style. Full instructions for the setup of macros will be covered in the software's help file or online wiki.

Waterfall Displays

FFT (Fast Fourier Transform) displays, also known as 'Waterfall' displays, have become the standard tuning indicator for the vast majority of data modes software. The waterfall is simply a spectrum display that shows the signal level as different brightness or colour dots. Each sample produces a single line of dots and subsequent lines follow behind, to create a scrolling display. The reason this display makes such a good tuning indicator and signal detector is that it shows a historical record of what's been received over recent seconds or minutes. Even if a station has stopped transmitting, you can click on the historical trace to set the correct tuning point. For popular modes like PSK31 or FT8, a spectrum display will show the entire band's activity in one sweep, so it becomes very easy to spot new signals and to move around the band segment.

Soundcard Troubleshooting

Most of the problems encountered by newcomers to data modes are associated with getting the audio from the rig to the software. The solution is usually very simple, but you need to adopt a logical fault finding process or you will soon get in a muddle.

The first step is to make sure your rig is switched-on and tuned to a band with some activity. The best place to start is the FT8 section of the 40 or 20m bands (7.074/14.074

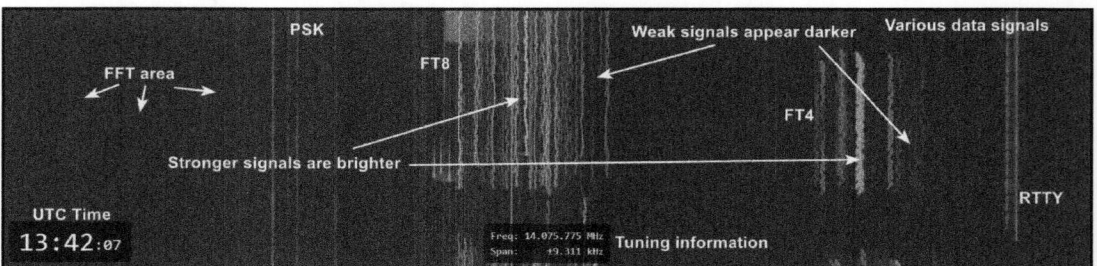

Fig 8.3: FFT Waterfall.

MHz). Make sure the mode is set to USB and that the bandwidth is set at its widest.

The cables carrying the audio data to and from the PC need to be stereo. Plug one end of the cable to the 'Line In' jack on the sound device that will be used for data modes. Failing that the microphone socket can be used but it is very easy to overdrive the input. The free end of that cable needs to be connected to, ideally, the fixed level audio output of the transceiver. Most, if not all have this facility but may require a consult of the instruction manual to locate. The fixed level output may be a jack socket, or may be a pin on an accessory socket (usually a DIN socket of some sort). For a quick test, the audio cable can be plugged into the transceiver headphone socket.

Open the data mode software and run through the configuration, ensuring you select the correct sound device and start the software decoding. Advance the transceiver volume control and you should begin to see the 'waterfall' brighten. Advance the volume control until the software input level is at the recommended level. Next, try clicking on a signal and, with any luck, you should be able to decode it. If the waterfall remains dark irrespective of volume level, the Windows sounds and/or mixer controls need to be investigated.

On most Windows 10 systems there is a speaker icon in the taskbar at the bottom right of the screen. Right-click and choose 'Sounds'. That will open-up the Sound panel with 'Playback', 'Recording', 'Sounds' and 'Communications' tabs. Select the 'Recording' tab. If all is well, you should see some activity in the bar graph display to the right of the device name. If the bar graph is dead, double-click on the device to open-up its Properties panel and choose 'Levels'. In this section make sure the slider is well advanced to the right and check that the speaker icon is not on mute. Close that panel and return to the 'Recording' tab. The data modes sound device should show one, or more, green bars. If the tests so far have worked and you

have signs of life in the 'Recording' tab but not on the waterfall, then you need to look at the software setup. The precise steps will depend on the software you're using, but you need to access the configuration and make sure you have selected the correct sound card and input. If this fails I suggest you visit

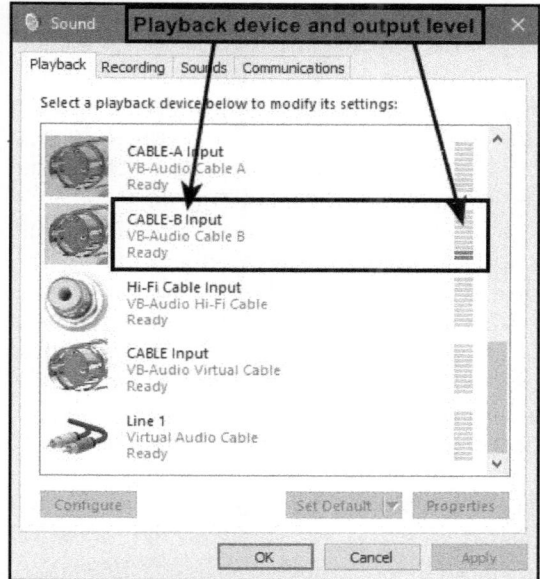

Fig 8.4: Playback Control

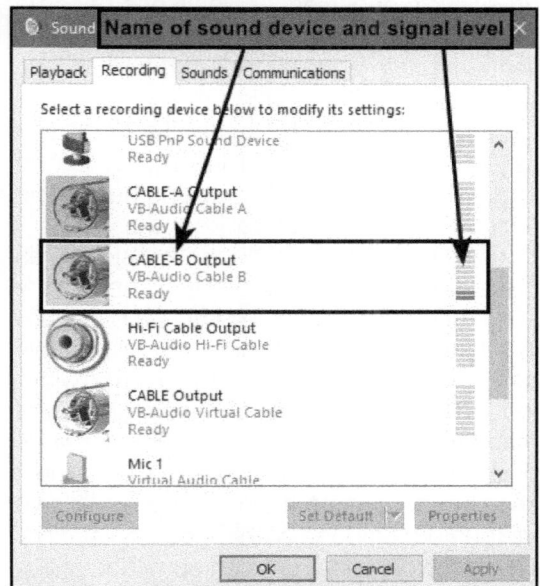

Fig 8.5 Recording Control.

the support forum for your software to see if the users have any suggestions.

Windows 10 updates have been known to cause sound related issues, and it may be necessary to check that the correct audio devices are selected and have the appropriate permissions granted. This has been discussed many times by the online communities of data mode users so finding an answer to your own issues should be fairly simple. Once everything is running correctly, any temporary connections can be tidied up and made permanent. Don't forget to put some good quality ferrite chokes on all the cables going to and coming from the PC/laptop. It may not be necessary in all cases but it is better to do it now rather than try and do it at a later date when everything has been installed. Any audio breakthrough issues will not become apparent until all bands have been transmitted on at the desired power level, so investing in a quantity of ferrite clip-ons or toroid cores will be money well spent.

Transmitter Drive Level

PSK31 was the most popular data mode for some time. More recently, FT8/FT4 has taken over that mantle, and with it the subject of drive levels and 'clean' signals has become a hot topic of discussion. A tune across the various data mode sections of the HF bands will show a wide variation in the quality of transmitted signals. Most are of a high quality, but a few will be overdriven or distorted spreading across a disproportionally wide section of the receiver passband.

Transmitted signals become distorted as a result of having a higher than necessary audio drive level from the computer overloading the transceiver's audio input. When this occurs the output signals can be much wider than they should be, due to the inclusion of many, usually suppressed, carrier sidebands plus harmonics of the original audio signal showing as phantom signals up and down the audio passband of the receiving station. The received spectrum can sometimes display signals that look to be over driven but are

in fact due to inadequacies within the receiver. If you find a distorted, dirty looking signal, try adding some attenuation in the receive signal path, it may improve the signal quality.

If the output from the computer is at the recommended level, the transmitter should remain within its linear region thus giving a good, clean, transmitted signal. All the software discussed here uses the audio output from the PC's soundcard to drive the transmitter. Because of that it's important to be able to adjust the audio output level, so as not to overdrive the transmitter. Whilst some data modes do not require the transmitter to operate in linear mode, most others do. For beginners it always recommended to run the transmitter in linear mode with a good, clean input signal. A simple way to ensure this is to switch your data modes program to transmit and adjust the audio output level so that the ALC (Automatic Level Control) just starts to kick in on the transmitter, then just back off the drive slightly until the point where no ALC deflection is shown. The way in which ALC is applied may be different in some transceiver designs and it may not be possible to achieve a zero ALC indication. Consult the manual that came with your radio, and/or the online community, as to the proper adjustments for digital mode operation. You may find that the required drive level varies between mode and band, so it is helpful to have an easily accessible gain control potentiometer in the transmitted audio path. This can be a home-made arrangement (using well screened leads), or built into your interface unit. Whilst most data modes software includes a facility to adjust the transmitted audio level, a manual control is faster and more convenient.

Mulitmode Data Decoders

Introduction

There are many very capable multimode decoding programmes/packages available, some free, some not. Some of these programmes are part of a larger software suite that interact with each other to form a complete integrated system. It should be noted that most of the software decoders available require Windows 7, or newer, to run. Some may work under Windows XP, but these are the exception.

It would be a very difficult job to list every programme and every mode covered by them. The software in this chapter has been chosen for a number of reasons: 1) Popularity – the more popular a programme is, the larger the user/experience base is; 2) Modes covered. In order for a data mode programme to remain popular it must be capable of supporting the latest/most used modes; 3) Performance. The software must be capable of providing good results across a wide range of hardware – both transceiver/receiver and computer; 4) Development. It is important the software is actively supported by its author(s) so it remains usable under the newest operating systems and reported bugs are tackled.

Fldigi

Introduction
Created by W1HKJ, Fldigi is a very versatile multimode decoding system, consisting of various programme modules including, amongst others, Logging; DX Cluster; Rig control; CW keying and HF messaging, in addition to the decoder module itself. There are versions of Fldigi for most platforms, including macOS (Mac OS X) and Raspberry Pi (Rpi).

Installation
For Windows, clicking the link on the home screen will open the download page, where various versions of the files are listed. These files can also be downloaded from the Fldigi project page on the 'SourceForge' website (sourceforge.net/projects/Fldigi/files). The Windows file will take form of 'Fldigi-4.1.20_setup.exe' (where 4.1.20 is the version number, changing as the software is updated). The download size is only 6.3MB, which is very small compared to some de-

coder installation files. Save the installation file to a convenient location, such as the default Windows 'downloads' folder. Once downloaded, double click the '.exe' file to begin the installation. If you accept the default options during installation, the software will install in the 'Program Files' directory with desktop and toolbar links to launch the software. When Fldigi is installed, 'Flarq' is also installed. Flarq is a client programme allowing the transfer of messages/mail or small files over the air.

Configuration
Running Fldigi for the first time will open the configuration wizard that guides you through the most important settings, enabling you to get on the air quickly. Firstly, enter your call sign, name and station details. It's worth thinking about this because the data stored here will be used in the transmit macros that you'll be using to save typing whilst on the air. Next, select the audio input and output settings and, for a Windows system, tick 'PortAudio' and choose the recording and playback sound cards that you intend to use. The Settings and Right Channel tabs can be left at their default settings. The final step of the configuration wizard is to set up Rig Control. If you don't use a rig control system, you can ignore this panel. Users of rig control systems will need to refer to their rig control instructions to establish the correct settings for this panel. That completes the basic configuration, but if you want to make any changes at a later time, all of Fldigi's configuration settings are accessible via 'Configure' on the main menu bar.

Operation
Fldigi supports a wide range of data modes, some of which are unique to this software. Considering the versatility of Fldigi, the interface has been kept simple but, with clever use of sub menus and mouse right-clicks, virtually everything can be customised to suit the operator. At the very top, mode selection is done via the 'OpMode' menu item. The

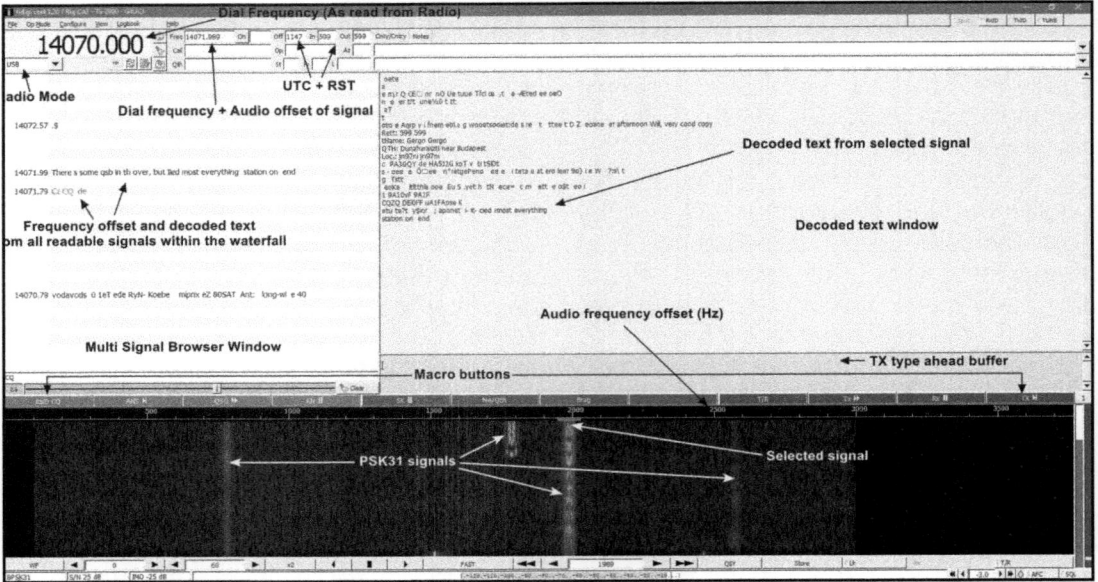

Fig 8.6: Fldigi layout.

section immediately below the menu bar is a logbook entry panel, with entry fields available for pertinent information, such as call sign, operator name, QTH, signal reports in and out, etc. At the end of the QSO press the 'save' button to store that information in the log. The 'save' button is the lower of the stack of three buttons to the left of the log.

Immediately below the log is the received text panel. This is where all the decoded text messages are displayed. You will also find your transmitted text shown in this section after it has been sent, but it will be coloured red so it's easy to identify. This panel effectively keeps a full running log of all your QSOs in the current operating session.

The pale blue panel below is what's known as the 'type-ahead' buffer. This is where text messages can be created and stored prior to transmission. The contents of this section will be transmitted as soon as you switch to transmit.

Immediately below the type-ahead buffer is a line of macro buttons. These are pre-configured with sample macros that you can edit to suit your operating style. Editing these macros is very simple – just right click on a button and the macro editor will open. Fldigi

has a powerful set of macro commands and it is worth taking time to read the excellent user guide to make the most of this feature. However, to get on the air quickly it is very simple to customise the default macros. Once the macros have been updated go to 'File – Macros – Save' and either overwrite the existing file or save with a new name. When you re-start the programme, you will need to re-load those saved macros.

The next section is the all-important waterfall/FFT spectrum display, where you can view all the available signals within the filter passband, then select a signal and begin decoding just by a simple left-click. Finally there is a set of controls for the waterfall display, transmit/receive switching, etc.

Ham Radio Deluxe – DM780

Introduction

Simon Brown, G4ELI, created the well-known and respected Ham Radio Deluxe (HRD) suite of programmes, which comprised of various modules for logging, radio control, etc., plus a data modes decoder, called DM780. HRD was free to use and became very popular, so much so that it was this author's 'go to' software for datamodes and logging for a considerable time. The last free versions of HRD were v5.xx. HRD v5 is not compatible with Windows 10, even when launched in 'compatibility mode'. Simon sold the software source code, etc. to a group of American amateurs, who took on the task of updating and expanding HRD. Starting with the new v6, HRD evolved and offered more comprehensive logging, radio control and an improved DM780. It should be noted that HRD is no longer freeware, but it is available as a 30 day fully functioning download which should give ample time to see if the software is suitable for your operating style.

Installation

Although this review is just for DM780, it is necessary to download the full HRD installation file, as it is not available as a standalone programme. From the HRD home page, click the 'Download' button, where the latest versions are available. The download is around 115MB, so bear this in mind if you have a limited data allowance. Once downloaded, double click on the .exe file to run the installer. You will need to tick the licence agreement before installation can continue. Installation takes around 30 seconds with a fairly recent computer. It is possible that HRD will need to download extra files to complete the installation. Should this be the case, HRD will notify and ask for consent before downloading anything further. Windows 10 or 11 will likely have all the prerequisite libraries installed but Windows 7/8 may not.

Configuration

When installation is complete and HRD is run for the first time, the 'Licence Key Manager' screen is shown. For evaluation purposes, click on the 'Register for a 30 day free trial' link. Once completed, the key will be emailed to you and should be entered in the appropriate box. The tick box agreeing to data collection must also be ticked, else the software cannot be activated. Once activated, the 'Software Expiration' and 'Licence Type' boxes will auto fill.

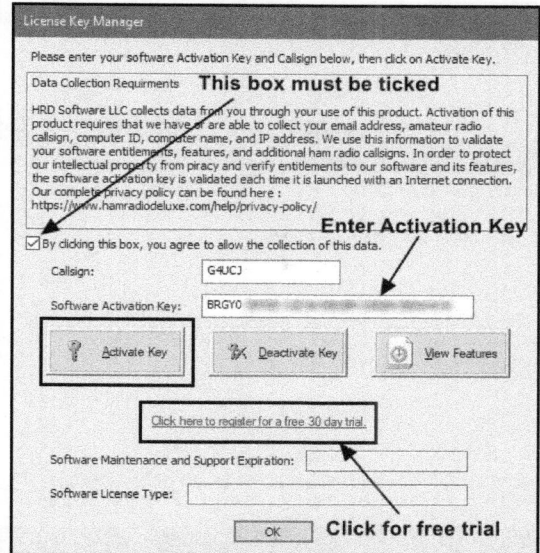

Fig 8.7: **Licence management screen.**

Once HRD is running, there are options to set up rig control. For the review, HRD was connected to SDR Console v3 (another programme by Simon Brown, this one being for SDR control) via a virtual com port pair and using the protocol for the Kenwood TS-2000. HRD can now tune the SDR and will also respond to commands sent from SDR Console (such as frequency, mode or filter changes). DM780 requires various parameters to be set in order to get the best from the software. If you have used HRD v5.xx then the options and overall 'feel' will be very familiar (even to the point where the callsigns used in some examples remain Simon's). Work through and adjust the options to your operating preferences. The most important option is sound device selection. Without the correct device selected, DM780 will be unable to decode any signals. The DM780 waterfall can be set to accommodate a visible bandwidth of 3900Hz. Pre-DSP transceivers typically have ceramic or crystal filters fixed at around 2.4 kHz for SSB. DSP and SDR transceivers have variable filters that can be adjusted to around 4 kHz wide. Using the widest filter settings will allow the greatest number of signals to be visible for decoding.

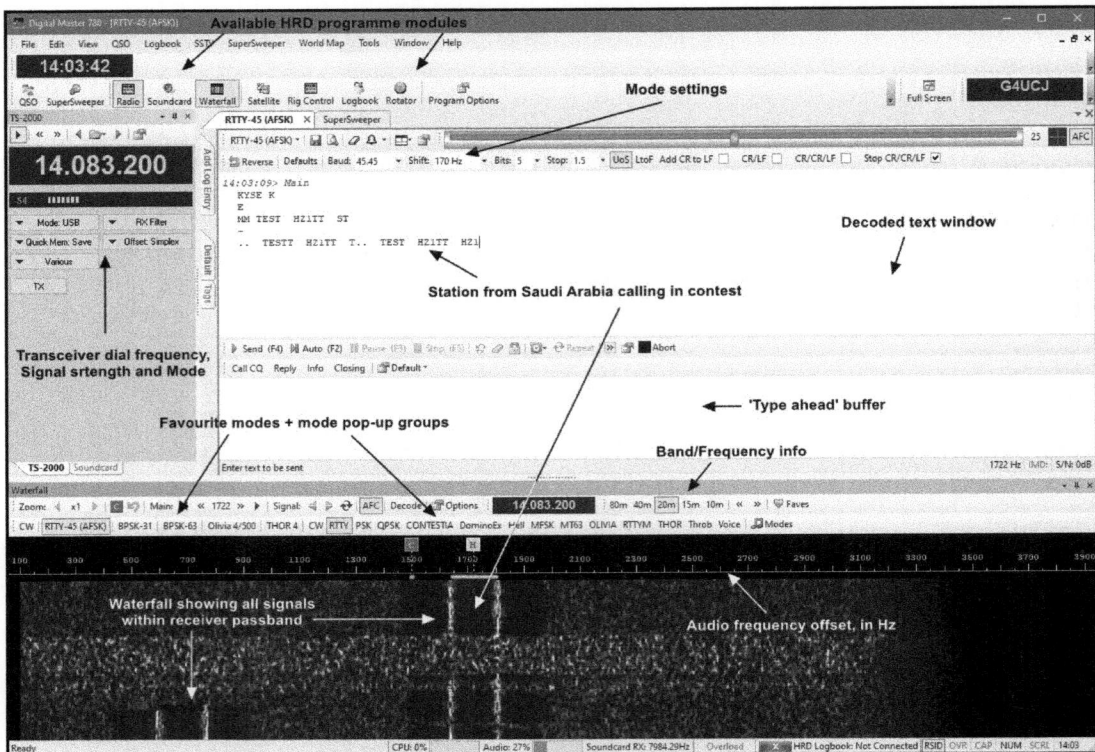

Fig 8.8: DM780 main decoding screen.

Operation

In addition to using the menu immediately above the waterfall display, the operating mode can be selected using the drop-down menu at the top of the TX/RX text box. Once the desired mode and frequency has been selected, any activity will be shown on the main waterfall display. Tuning to a station is simply a question of clicking the mouse in the centre of the signal. The bandwidth of the signal is shown by the highlighted bar that follows the mouse click. Received text appears in the top text box, as does the transmitted text (after it has been sent). The lower text box is the transmitting 'type ahead' buffer, where you can prepare messages ready for transmission. The combination of DM780, with HRD's logging and control functions make for a very powerful and sophisticated setup that is beyond the scope of this brief overview.

MixW

Introduction

MixW is a commercial multimode programme written by Nick Fedoseev, UT2UZ and Denis Nechitailov, UU9JDR. Originally a DOS programme, the first versions of what was to become MixW (v1.xx) were written back in the early 1990s. Windows versions of the programme (MixWin as it was then known) came to being in 1998. The last release of MixWin supported just 5 modes (3 'proper' data modes: PSK31, RTTY and Packet, plus a CW decoder and support for SSB logging). Version 2 introduced a completely new GUI (Graphical User Interface), many improvements over the previous version and the name changed to MixW. Version 3 introduced comprehensive networking facilities, allowing automatic linking to the DXCluster network, plus the ability to store logs on remote servers. Version 4 of MixW is a complete redesign, once again using the latest software innovations to improve decoding accuracy.

All versions are 'paid-for' software, but they have a fully functional 15-day trial period. The trial allows the user to 'try before they buy'. The software will continue to work after the initial trial period, but a 'nag screen' appears after every mode change (and possibly at other times) staying on screen a number of seconds before allowing the selected mode to be used. Below are descriptions of both v3 and v4. There are so many differences between the two versions that it was thought necessary to show them separately. To cover the widest range of modes, both versions are necessary and they can be installed and run on the same computer without problems.

MixW 3

Installation
Navigate to the 'Download' page and choose the MixW version you wish to install. The MixW 1.45 download is a mere 664kB, increasing in size up 13MB for the MixW 3.2 file. The download consists of a single executable file (SetupMixW3_2.exe or similar). Installation is simply a matter of double clicking the downloaded file and following the on screen instructions.

Configuration
When MixW is started for the first time, a panel will open so the user may enter personal data, such as callsign and station details, which are then used macros and logs. The next step is to make sure the correct sound card and interface has been selected. To do this, choose Configure – Sound Device Settings from the Menu bar and use the drop-down selection to choose the correct sound card. If you are using a proprietary interface it's worth using the separate Interface wizard to help complete the configuration. The wizard can be found via the menu at Configure – Interface Setup Wizard. Immediately below the menu bar you will find a

set of Macro buttons that can be configured using a right-click.

Operation
MixW is very easy to use and, like most of the programmes here, uses a waterfall display to show activity and aid tuning. To receive a new station you simply click the mouse on the waterfall trace and decoding will start immediately. MixW also makes use of your Internet connection to automatically look-up received call signs, report their location and even tell you whether or not you've worked them before. Double clicking on a call sign in the received text box will automatically transfer that call to the electronic log and the stored call will also be used in the macros. If you get in a muddle, the Escape key causes an immediate switch back to receive.

The screenshots **Fig. 8.9 and 8.10** show MixW3 receiving two different modes (CW and PSK63). At the time these were taken, a contest was active on both modes, so there was plenty of activity. The MixW screen offers various information at a glance, such as operating mode and associated settings including speed/shift/offsets, etc. along with system time and date. Along the top of the screen are the macro buttons, 12 at a time, corresponding to the appropriate Function keys (F1-F12), each having the macro name shown. Pressing the 'Shift', 'Ctrl' or 'Shift + Ctrl together' keys brings up other sets of macro keys, with a total of 48 being available.

A partial logbook can be shown on the screen, with the full log and search facilities available at the press of a button. The main receive screen can be configured to suit your needs. Options such as font, background/foreground colour can be assigned. Callsigns can be configured to show their logbook status, a callsign can be shown in a different colour to signify if it is a new callsign, a new DXCC, new prefix (or multiplier if in contest mode) or if that callsign has been worked before. If bookmarks are selected, a mark will be placed in the waterfall area to indicate a callsign has been decoded at

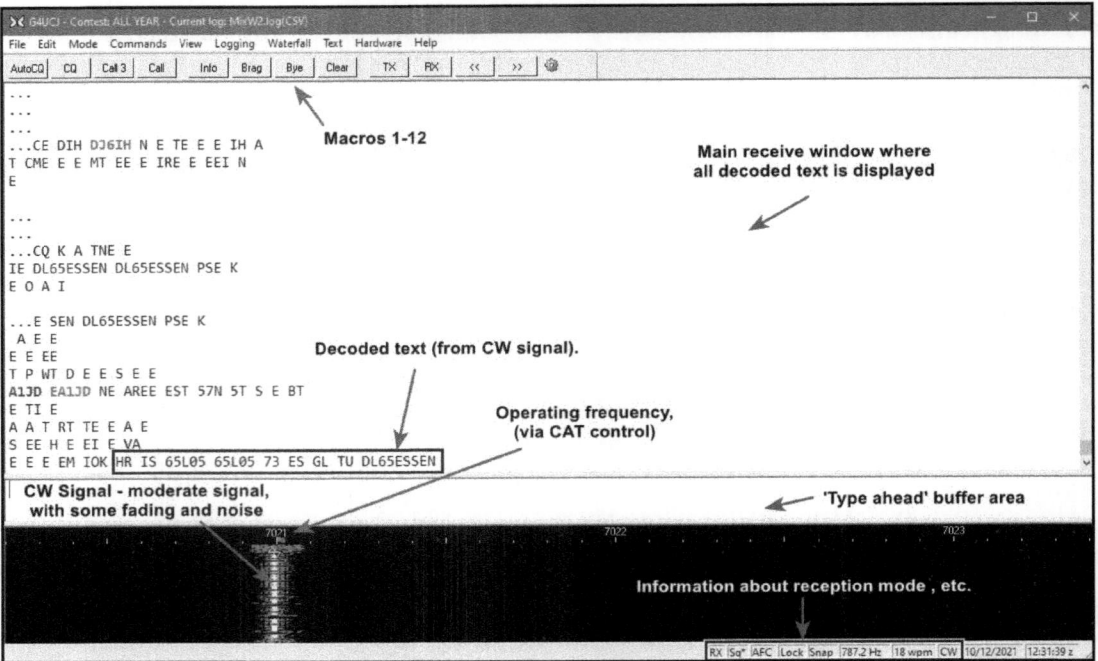

Fig 8.9: MixW 3 decoding CW.

that frequency. This is a super feature which allows the user to avoid duplicates as the marker will be in a different colour to the mark of an unworked station. The bookmarks will also indicate DX spots on the waterfall.

In **Fig 8.9** MixW is decoding a CW signal from a special event station in Germany (DL65ESSEN) on the 40m band. The CW image was taken with the receiver using a very narrow filter bandwidth (80Hz in this case), whereas the PSK63 image was taken with a wide (4kHz) filter selected. The wide filter enables the entire passband, and any signals within it, to be seen. The receiver in use was an Airspy HF+ Discovery SDR, but a similar display would be seen if using the audio output of an analogue receiver with conventional filters. Although MixW 3 is not a dedicated contest logging

programme, it does have the facility to define various contests and calculate scores by using specific contest dll files (these are small additional files that allow the score to be calculated correctly using log entries in conjunction with a file that defines the multipliers for the contest in question (i.e. CQ zones, US States, DXCC, etc.). The contest dll's can be downloaded from the MixW

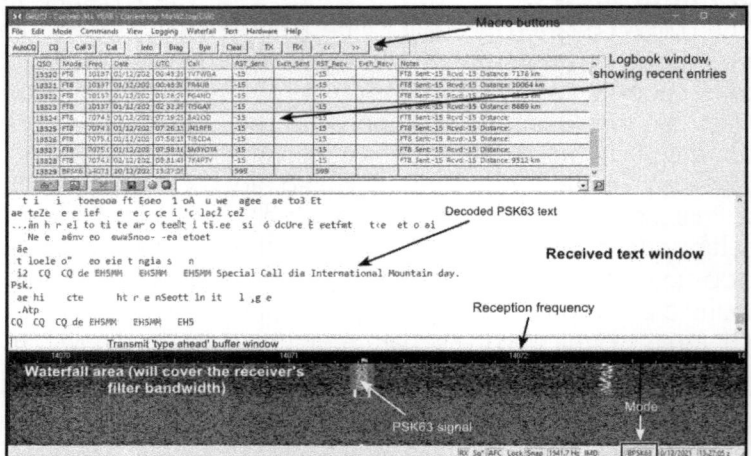

Fig 8.10: MixW 3 decoding PSK63 signals.

website. **Fig 8.10** is a more detailed image showing the logbook and a special event station from Spain (EH5MM) being received on PSK63. Notice the difference in the waterfall between the two images.

MixW 4

Introduction

MixW 4 is the latest incarnation of the well-known multimode decoding software. Building on the features found in MixW3, this new version has a more modern, improved, look to the GUI; fully configurable contest capability; data modes decoding, including FT4/FT8 and logging facilities. The software is regularly updated, with the authors fixing bugs and adding features.

Installation

Double clicking the downloaded setup file begins the installation process, which is similar to the one used by the majority of Windows software. If the default paths are followed, the software will install and function perfectly well. There is the option to install to a folder of your choice, should that be desired.

Once installed, there is a 15 day trial period for the software. This should be sufficient to find out if the software suits your needs. The website informs the software will remain fully functional after those initial 15 days have elapsed, however this may mean the inclusion of a time delayed 'nag screen' appearing after certain things, such as modes are changed. After the 15 day trial has expired you are presented with the choice of purchasing the software in full or, if you have a previous version installed, you may purchase an up-

grade licence. All versions of the full MixW licence are the same cost (£45/$54/€53). The upgrade licence costs £15/$18/€17.50 – prices are subject to exchange rate fluctuations. Once registered, you will receive a file via email that will need to be copied into the MixW folder (instructions will be enclosed in the email). The next time MixW is opened, it will be registered and any restrictions/nag screens removed.

Configuration

Once installed, you will need to enter your callsign, QTH, grid locator, etc. Work through each screen, in conjunction with the instruction manual to ensure everything is configured before you begin to operate.

Operation

I judge software, initially, on how user friendly it is – can I start using the software without having to refer to the user manual every few minutes? I used MixW 4 without looking at the manual, but became stumped when things

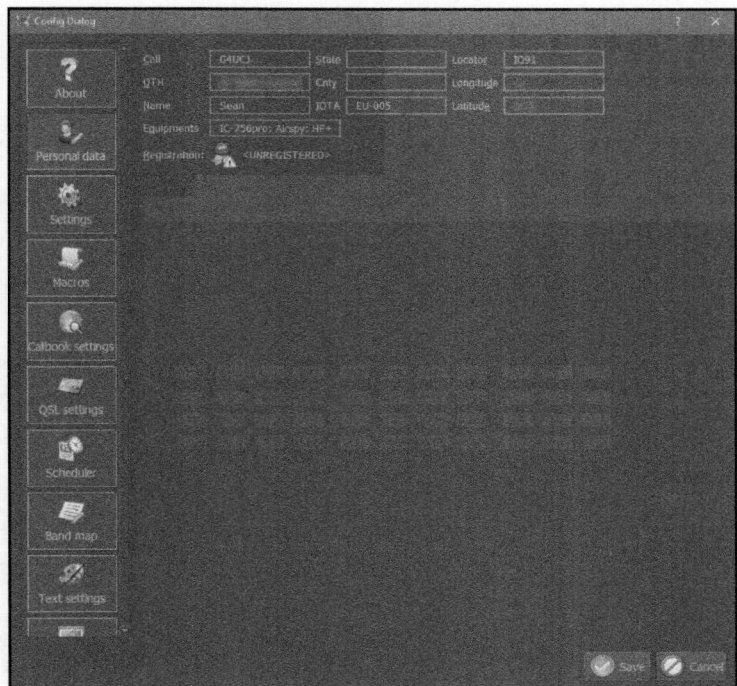

Fig 8.11: MixW4 configuration.

would not do what I wanted them to. The manual for this software is a vital tool if you are to get the most out of it. I got the software working without referring to the manual, but the hit and miss approach will only take you so far. On the surface the software looks simple but there are lots of functions that are hidden away which are only revealed by luck, or consultation of the manual. The screen layout of MixW 4 can be frustrating, initially, as the moveable windows do not act in the manner that you may expect and it is easy to end up with a full screen window of one part of the software obscuring others. The bottom left corner of each window has a 'dock/undock' button which is the secret to setting the layout to your liking, but it is still easy to make a mistake with the window sizing and having to go back a step to undo the error.

The CW decoder in MixW has always been rather good (particularly using the v1.45 algorithm) and MixW 4 continues in the same vein. Using an indoor loop, with moderate signals on a noisy band, the decoder did a surprisingly good job. As the signal becomes noisier, the decoder accuracy obviously reduces. No matter how good, CW decoders are no match

for human ear/brain combination, which is the best decoder there is. The CW decoder works best with machine sent code, such as used during most contests. In this scenario, the CW decoder performs very well indeed.

Macros

MixW4 has 4 blocks of 12 configurable macro buttons, giving 48 in total. To use macros 1-12, press the corresponding Function or 'F' key, (F1-F12). Macros from 13-24 requires pressing the 'Ctrl' button plus the relevant 'F' key. Macros 25-36, pressing 'Shift' + 'F' keys, and macros 37-48 by pressing 'Ctrl', 'Shift' & 'F' keys together.

The main screen of MixW 4 is fully configurable to suit you operating style. All windows can be moved, resized and docked. Initially, it took a while to become accustomed to the way the windows interacted. The logging window is split into two parts, one for data entry of the current QSO and the second window displays the rest of the logbook. **Fig. 8.12** and **Fig. 8.13** are two layouts. These layouts can be saved and recalled as desired. There are a number of windows and a great deal of data that can be displayed which would

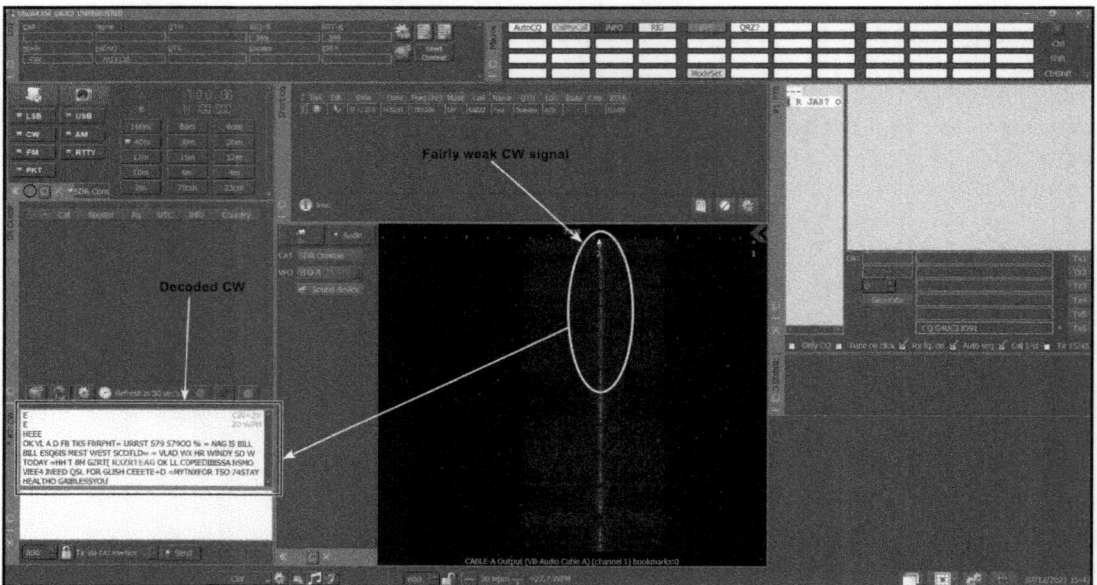

Fairly weak CW signal

Decoded CW

Fig 8.12: MixW 4 decoding CW.

suit a very high resolution monitor, such as a 4k capable unit. These layouts were used on a 24" full HD monitor and some windows were hidden and data hidden from view underneath. It may take some time to find the best layout for your style. Maybe a layout for each mode would be the way to go. As yet this has not been pursued. **Fig. 8.13** is an image of MixW decoding numerous FT8 signals and the layout has been adjusted to show the windows that are useful to this mode. BY having several layouts saved, it is possible to have custom screen layouts for each mode.

MultiPSK

Introduction

MultiPSK is a true multimode decoding programme, written by French amateur Patrick Lindecker, F6CTE. It features an extremely wide range of modes that include: PSK, CW, Packet, PACTOR 1, AMTOR, ASCII, MFSK8, MFSK16, Olivia, Contestia, Throb, Domino, PAX, ALE, JT65, FELD-HELL, HF-FAX and SSTV. In addition, MultiPSK includes receive only capability for a number of commercial and specialised

modes, but there is a time limit of 5 minutes per session when using an unregistered copy of the software to decode these modes. Currently, it costs €35 (approximately £29) to register the software. Once registered, a small file is emailed to the user and this is placed into the main MultiPSK folder. After restarting the users name and address will appear on the configuration screen, indicating the software is now registered and free from restrictions. Note that the registered address cannot be changed. Whilst not an issue as far as operability is concerned, it is a little frustrating not being able to change an address that is over 10 years out-of-date and emblazoned on the programme.

The inclusion of so many modes and options has made the GUI extremely crowded, which can be confusing for the user and takes quite a time to get used to. Persistence is rewarded though as the programme is packed full to the brim with features and signal analysis tools that are not usually found outside of professional software costing hundreds or even thousands of pounds. When observing a mode that is unfamiliar, MultiPSK tends to be the first tool used. I purchased MultiPSK about 15 years ago and it is still in regular use because it covers

Fig 8.13: MixW 4 with alternate layout.

those unusual modes that are not found on other decoders. MultiPSK is under constant development with new modes added from user requests and bugs/performance improvements are made as a result of user feedback.

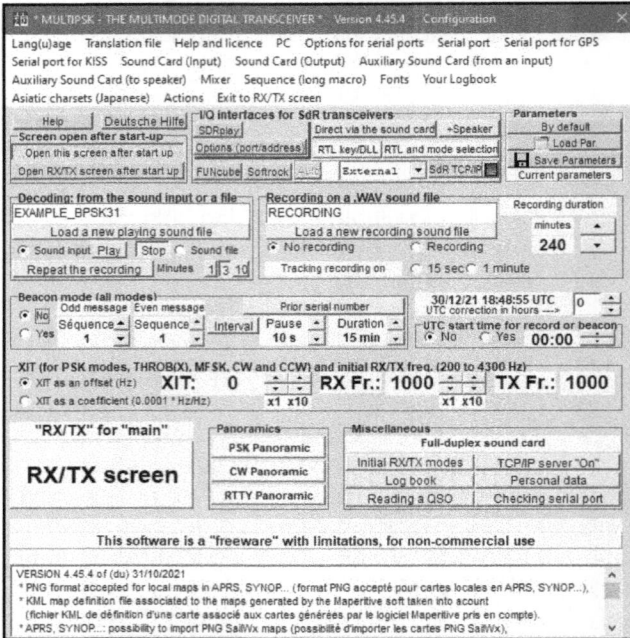

Fig 8.14: MultiPSK configuration.

Installation

The software setup file is around 16MB. Once downloaded, double click the MULTIPSK_ setup.exe file to begin installation and follow the instructions. The installation process has improved enormously over time.

Configuration

The initial screen shown at start-up is the configuration panel and the first task is to run through the top menu and set the 'Sound Card (Input)' and 'Sound Card (Output)'. To get to the main operational screen of MultiPSK, press the large 'RX/TX Screen' button towards the bottom left of the config-uration screen. Macros are located just below the waterfall display on the main screen, and are edited via a right-click of the mouse.

Operation

In addition to a 'busy' interface, which harks back to the early days of Windows, general operation of this software can be a little quirky. To tune to a RTTY or similar signal you must

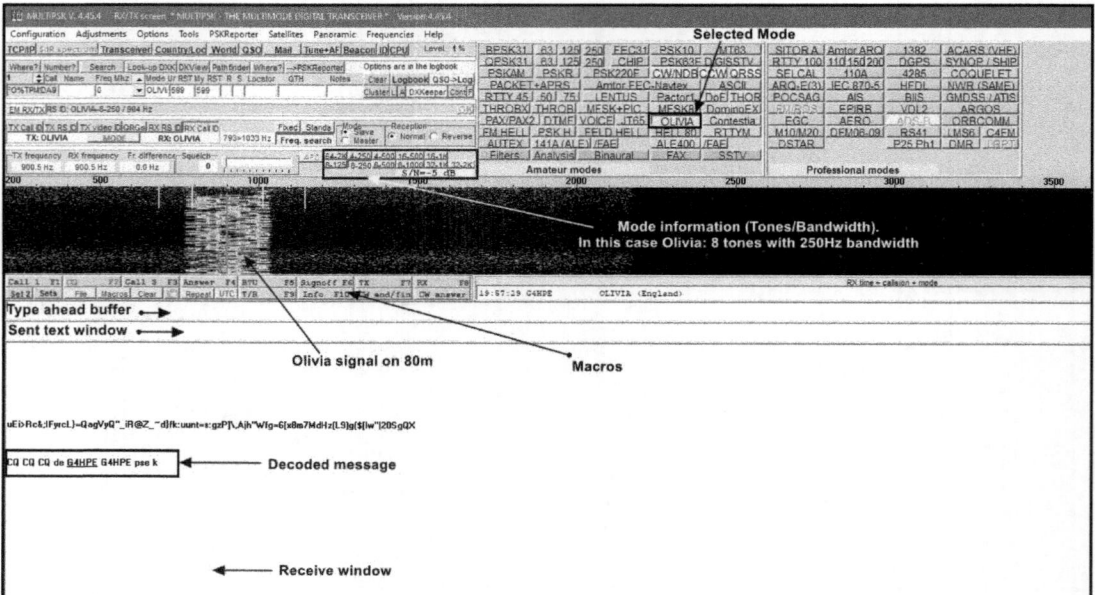

Fig 8.15: MultiPSK decoding screen.

click on the right-hand peak rather than the centre. The text screen layout is also different, with the received text occupying the lower section while the type-ahead buffer uses the upper text box. The central text box is used to show text after it has been transmitted. MultiPSK retains the use of the 'Escape (Esc)' key to abort transmission. MultiPSK can identify some modes automatically using the Reed-Solomon identifier system (RS ID). RS ID consists of a short data burst containing mode, speed and shift details, allowing the receiving software to be configured correctly, without operator input. The RS ID is sent just before the main data transmission commences.

The term 'Swiss Army Knife' is used quite a lot on the internet to describe tools that are capable of doing multiple things. If I were to describe any software as a 'Swiss Army Knife', then it would be MultiPSK – I cannot think of any other data mode decoder available, outside of a professional setting, which is able to decode and examine such a vast array of modes and signals.

MultiPSK has a very useful feature called the 'Panoramic' display and is available for CW, PSK and RTTY modes. The panoramic display allows multiple signals (channels) to be decoded simultaneously. With a reasonably modern computer, the panoramic display can decode up to 36 different signals (these are referred to as 'channels' by the software author). **Fig 8.16** shows the panoramic display in action during a busy RTTY contest. Four different signals are being decoded simultaneously in this example, but there are some repeats due to the software decoding the same signal again, possibly due to the stop/start nature of contest signals triggering the decoder repeatedly at slightly different frequencies. The panoramic display is a useful tool that requires some time to get the best from it.

TrueTTY

Introduction

TrueTTY is written by Sergei Podstrigailo, UA9OV (DXsoft) and covers a number of the more popular data modes, including PSK; RTTY; ASCII; Amtor/FEC; (AKA Sitor B/ Navtex); MFSK; Packet and Sitor A. TrueTTY integrates with AALog, also released by DXsoft. All the products have a 30 day evaluation period so you can ensure the software is suitable before committing to purchase (currently $39, €35, £29). DXSoft offers an array of software for the radio amateur, including an electronic logging programme; voice keyer; rig control; CW reader and CW terminal; NAVTEX/Pactor/GMDSS/WEFAX decoder and more.

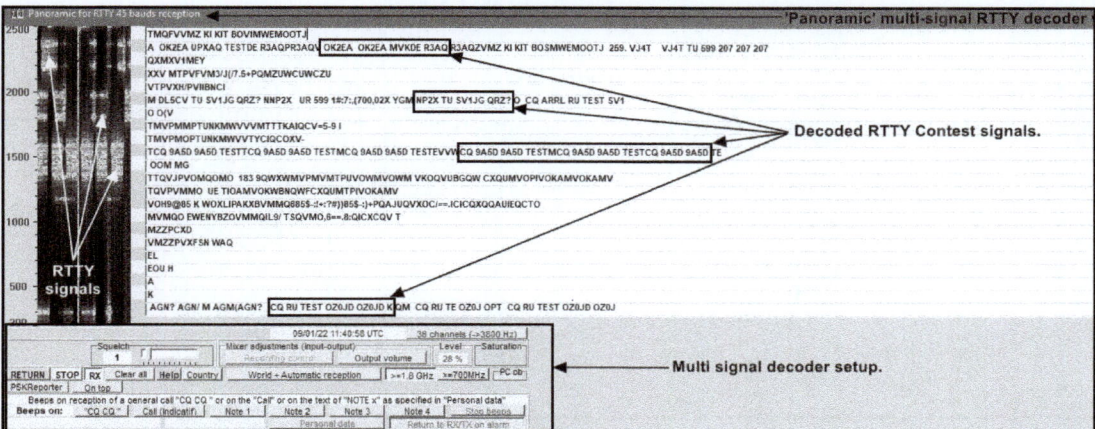

Fig 8.16: MultiPSK Panoramic decoder display.

Installation

The TrueTTY download is a zip file of 1.5MB and once downloaded should be extracted to a temporary folder. Once extracted, double click on Setup.exe to begin installation. For installation to proceed, the usual type of licence agreement needs to be accepted. Installation is quick and painless.

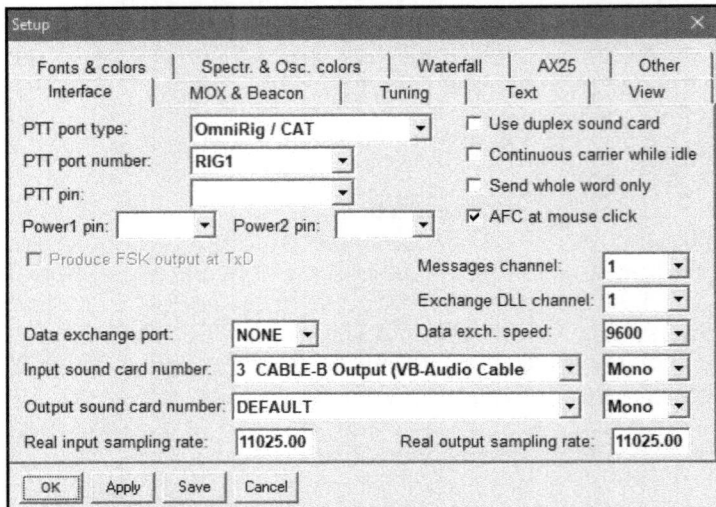

Fig 8.17: TrueTTY configuration.

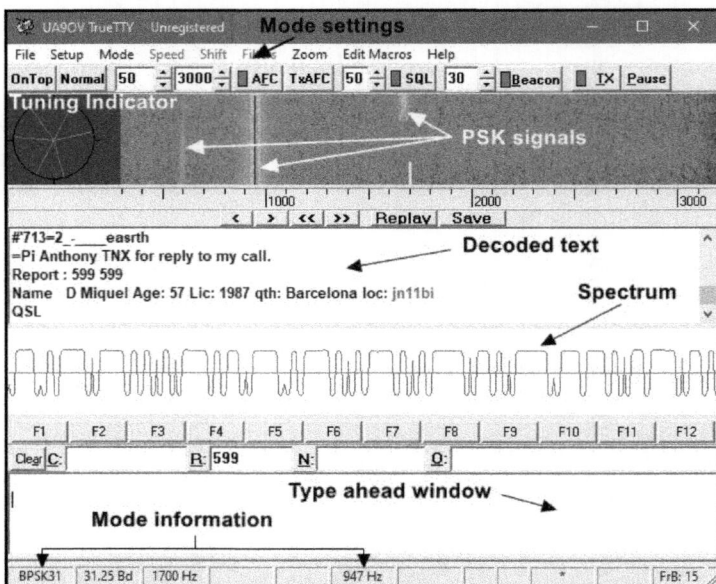

Fig 8.18: TrueTTY decoding window.

Configuration

The configuration screen of TrueTTY is quite straightforward and allows many parameters to be set/defined or changed. Having set a few important things, such as the correct sound device, the software was ready to use. I did not find it necessary to go through every setting in order to start using it, but it may be wise to consult the help file for assistance with some of the lesser used items.

Operation

TrueTTY has a compact interface that will fit onto the majority of screen resolutions and displays a useful amount of detail without looking overcrowded. 20m provided some PSK signals to test the decoding capabilities and the results were very good.

SeaTTY

Introduction

SeaTTY is another multimode decoder from DXSoft, this one concentrates on modes associated with weather and navigation warnings transmitted on HF and VHF, plus weather charts transmitted via fax (facsimile) on shortwave. As with other DXSoft programmes, SeaTTY has an initial 30 day evaluation period. To purchase a licence for the software the current cost is approximately $49/€39/£36.

Installation

As per TrueTTY.

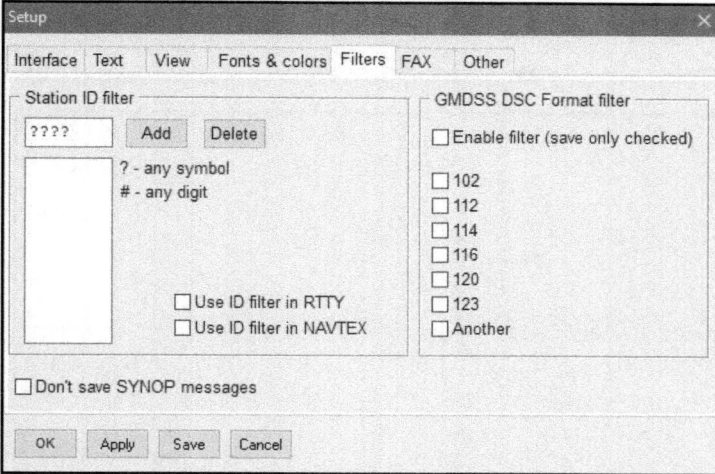

Fig 8.19: SeaTTY configuration.

Configuration

There are some specific to SeaTTY features than can be configured. A consult of the help file will be very useful to those not familiar with the terminology and use of modes relating to weather and maritime broadcasts.

Operation

If the user is familiar with TrueT-TY, they should have no issues using SeaTTY, as there are obvious similarities between the programmes. Having used SeaTTY for a couple of hours, the software was found to be quite forgiving with regard to input levels. Both high and low input levels were used and did not seem to affect the decoding.

The spectrum at the top of the screen enables the user to adjust the frequency so the black and/or white tones are aligned with the on-screen markers. SeaTTY allows both the slant and the horizontal position of the received image to be adjusted. If the soundcard has a slightly different sample rate than the system is expecting, the resulting image may be slanted one way or the other. To

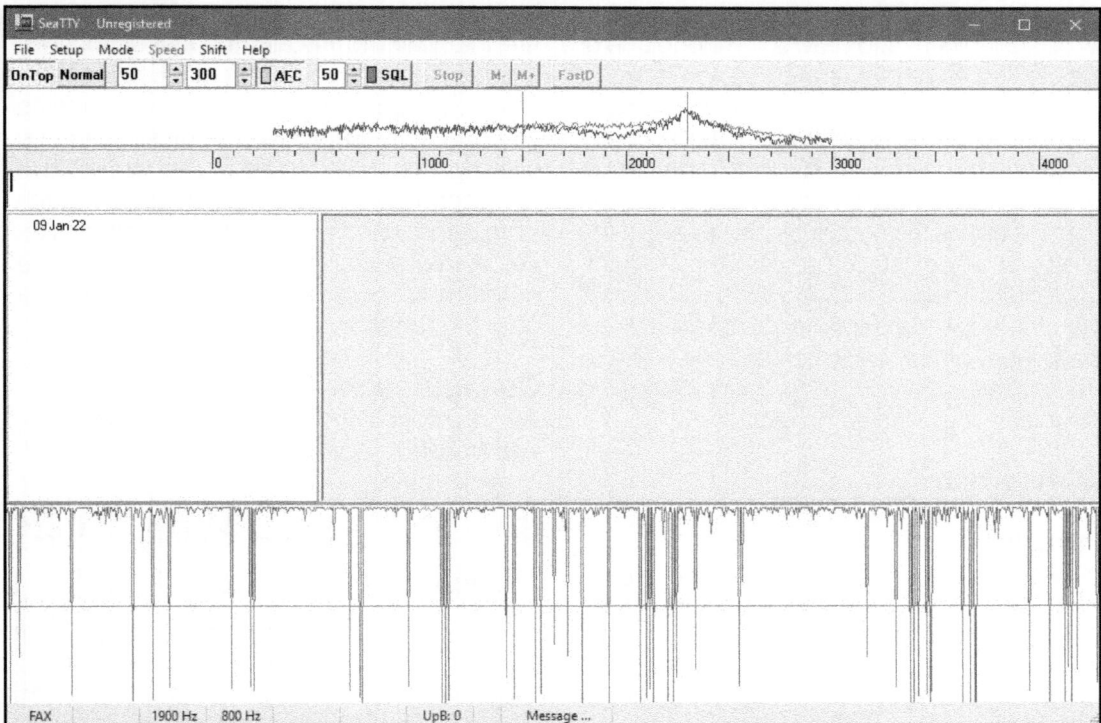

Fig 8.20: SeaTTY receive window.

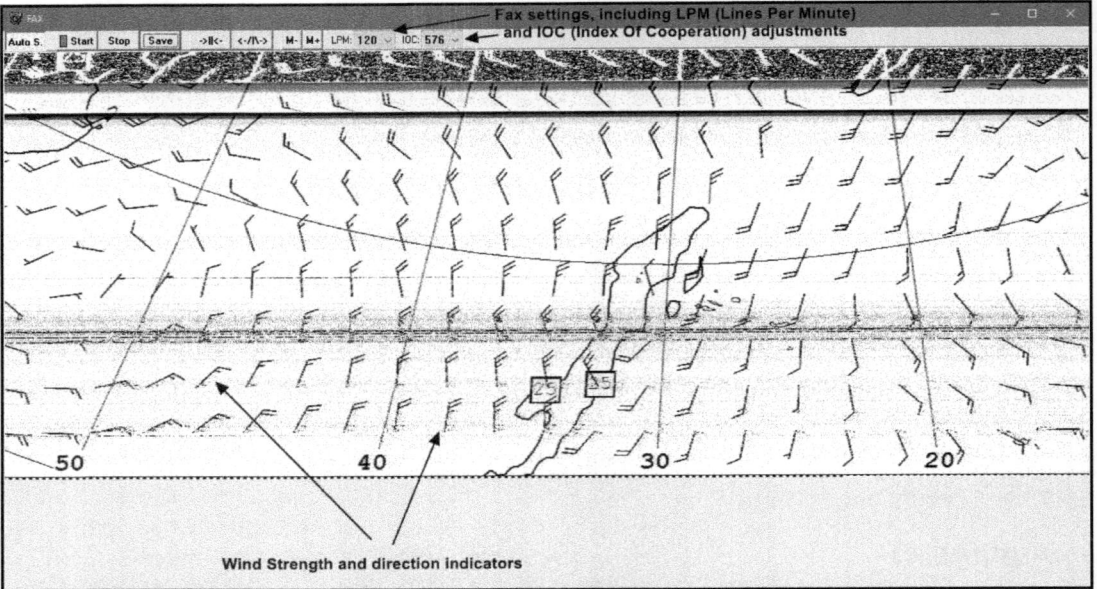

Fig 8.21: SeaTTY receiving WEFAX signal from Bracknell on 2.618.5 MHz.

correct this, wait until the sync pulse indicating the end of frame has been sent and the software has stopped recording. Next, click on the 6th button from left (next to the button showing two parallel lines) and draw a line along one of the slanted edge markers. Once you let go of the mouse, the software will correct the slant and display a message asking you to confirm the new sound device sampling frequency, which the user should agree to. The 5th button from left allows the horizontal placement of the image to be adjusted. The user may find the edge of the image is showing somewhere other than the edge, perhaps near to the middle of the image. Simply click on the edge line and the software will shift the edge marker ensuring the image is presented correctly.

WinWarbler

Introduction
WinWarbler is part of a set of eight interconnecting, free, programmes written by Dave Bernstein (AA6YQ) as part of his DXLab suite. WinWarbler is a datamodes decoder that supports CW, PSK and RTTY modes at a variety of speeds and shifts.

Installation
In order to use WinWarbler, it is necessary to install the DXLab 'Launcher' application, which is required for whichever part of the suite you are intending to install. Full instructions for this are documented on the DXLab website. Once the launcher is installed and running you can start the WinWarbler installation by clicking the 'ww' button in the launcher. If you accept all the default settings during the installation, WinWarbler will be installed in c:\DXLab. WinWarbler will run as a stand-alone programme, or can be integrated with some/all of the remaining DXLab software.

Configuration
As with most of the software described here, WinWarbler supports the use of transmit

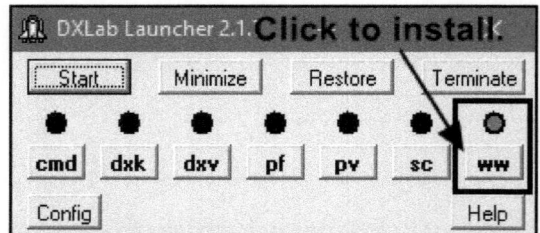

Fig 8.22: WinWarbler installation.

macros and a certain amount of automation of the log keeping process. One of the first requirements is to populate the software with your call sign and you may find that you are prompted for this when the programme starts for the first time. To update that information, click the 'Config' button in the top right-hand section of the main screen. When the Configuration screen opens you will find the call sign entry on the 'General' tab. The 'Soundcard' tab enables selection of the appropriate sound device. The Configuration screen provides access to a wide range of customization controls, including logging and contest operation. A selection of example transmit macros are provided in the macro section, which is located just below the text screens. The stored macro is displayed when you hover the mouse over the button and a right-click provides access to the macro editing screen, where all the macros can be customized.

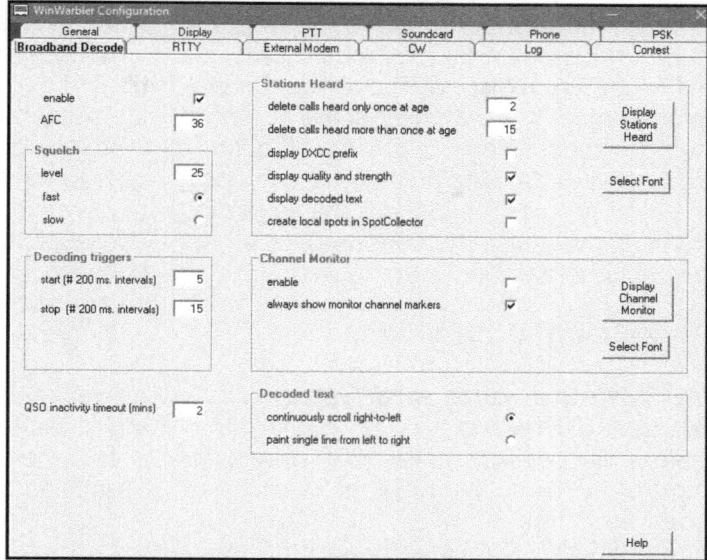

Fig 8.23: WinWarbler configuration.

Operation

The WinWarbler channel monitor can be particularly useful, as the software can decode up to 47 channels of PSK simultaneously. Activating the Broad-Band Decoder (BBD) is simply a case of ticking the 'BBD' box at the bottom left of the main screen. In addition to showing the decoded

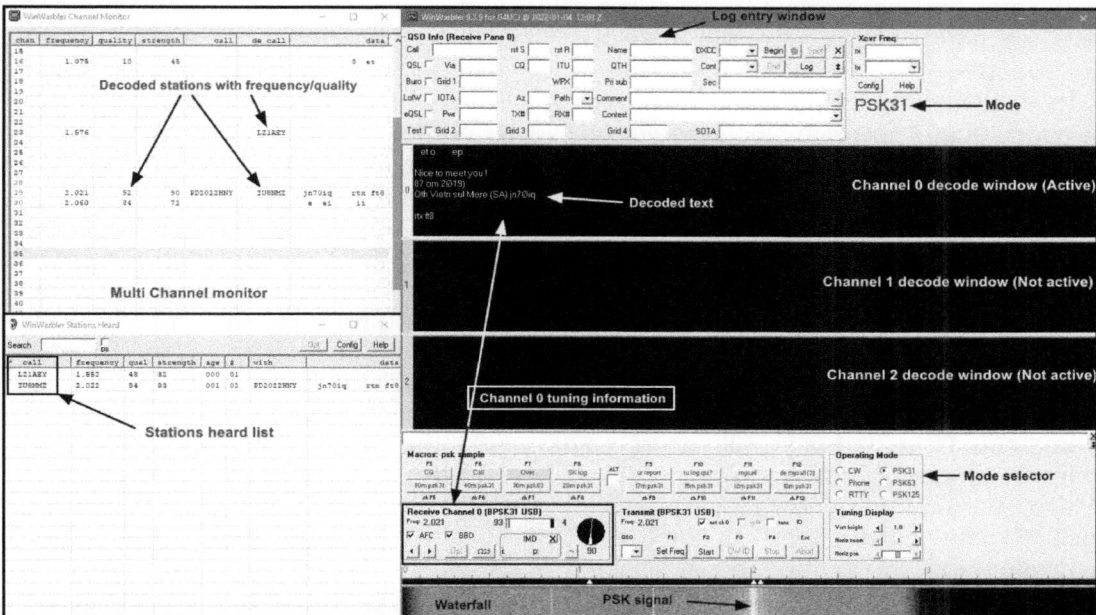

Fig 8.24: WinWarbler decode and multi-channel screens.

output in the 'Channel Monitor' screen, a number of small white triangles are visible in the spectrum display. These are channel indicators for the stations that are currently being processed by the 'Channel Monitor'. PSK activity has fallen off dramatically since the evolution of FT8, but can still be found, particularly during the various data mode contests held throughout the year.

'Specialist' Data Modes

PC Time Keeping and Data Modes Software

Before discussing software for 'specialist' modes, it would be prudent to discuss computer time-keeping and the in-built clock on the PC (or other computer). Although accurate enough for everyday activities and even some data modes, the on-board clock falls short in performance when time critical software are used. PC time keeping is far better than it used to be, but it is still not good enough for our needs. Windows synchronises the PC clock with its own internet time server once every seven days. This interval is far too long to be of practical use for most data modes. Even modern computers can have errors of ± a second or more over the course of a 24 hour period. This is fine for the average user, however the data modes enthusiast requires something with much greater accuracy. The simplest way to achieve this is by synchronising the PC clock to one of the internet time servers. To maintain accuracy, the clock should be synchronised every few hours. The synchronisation task has been automated by the use of freely available software. When setting up a time sync programme, a time server needs to be selected from a list. A time server in the same country as the user will ensure any internet latency is kept as low as possible. Make sure any documentation about the desired time server is read, as some have limitations on use or require permission/authorisation to use.

I use a small, free, programme called 'BktTime', written by the Italian amateur IZ2BKT. This programme loads at Windows start-up and can be configured to sync the PC clock at set intervals, from a few minutes to many hours or time can be manually synchronised. Whilst running, BktTime uses just 1MB of RAM so is very unlikely to have a noticeable effect on system responsiveness/performance. Other time sync software available includes 'Dimension 4'; 'Meinburg NTP'; 'Time-sync' and 'NetTime' as well as BktTime. Most issues with decoding software are due to PC clock accuracy, or lack thereof. If the PC clock is off sync by as little as 1 second, some decodes will be missed. Greater than 1 second and things start to deteriorate, rapidly.

WSJT/WSJT-X Suite

Introduction

The suite of free programmes written by Nobel Laureate, Professor Joe Taylor (K1JT) called 'WSJT' (WSJT is an acronym for Weak Signal (communication by) K1JT) has been developed and optimised for use with difficult/weak signal paths such as EME (Earth-Moon-Earth or 'moon-bounce'), ionospheric scatter, etc. Some modes are optimised for LF, MF and HF propagation whereas others are tailored to the VHF/UHF enthusiast, so it should be possible to find a mode for every kind of propagation encountered.

These modes are designed to allow a very basic QSO to take place, with just the minimum of information exchanged. Joe Taylor, along with his team of authors and testers, has provided excellent PDF manuals for all the software in the WSJT suite, and it is highly recommended these are used to help you get going.

There are two suites of programmes: WSJT-X and the older WSJT. Each programme is evaluated and an overall synopsis is made of the programmes that form the rest of the WSJT/WSJT-X download.

WSJT-X

Introduction

This is the newer, currently in-development software from K1JT, together with a development team including G4WJS (SK), K9AN and IV3NVW. The modes in this software are tailored to decoding fraction-of-a-second signals reflected from ionized meteor trails, plus steady-state signals that are >10 dB below the audible threshold such as those bounced off the moon, etc. WSJT-X supports the following modes: FT4; FT8; FST4; FST4S; JT4; JT9; JT65; Q65; MSK144 and WSPR.

Installation

The homepage offers installation files for Windows (32 & 64bit), Linux (Debian/Fedora 64bit and RPi (32/64 bit)) and macOS (10.13-12). The Windows 64 bit installation file for WSJT-X is around 31MB. Double click the downloaded .exe file (which will have a name such as 'wsjtx-2.5.2-win64.exe') and follow the on screen instructions.

Configuration

To run the software successfully for the first time you may need to run it with Administrator privileges. The next step is to choose 'Settings' from the 'File' menu item, then enter your call sign, grid locator (4 characters, e.g. 'IO91', is enough). Under the 'Audio' tab, select the appropriate sound device and then adjust the input and output audio levels, via the soundcard mixer or interface level controls. The input level should be adjusted so to show 30dB on the bar graph at the bottom left of the Spectrum screen.

Operation

When operating the various modes in WSJT-X, remember any QSO made will not be of the 'standard' type. QSO's

follow a strict sequence allowing only the required information to be exchanged, as efficiently as possible. A QSO in most of the WSJT-X modes consists of Callsign – grid locator – signal report (or 'R' + signal report) – 'RRR' (Received OK) – '73'. If the 'Auto Seq' box is ticked, the programme will advance through the QSO steps on receipt of the expected reply from the QSO partner. This is useful for MS (Meteor Scatter), where it is possible to be sending the same information repeatedly whilst awaiting a return from the other party, especially when reflections are scarce. This can take hours, so automated sequencing means you can take your eye off the screen without missing that one ping that contains the required data to continue or complete the QSO.

Operating Multiple Instances

It is possible to run two, or more, instances of WSJT-X, each instance will have a unique name, set by the user. To achieve this, firstly make a new shortcut on the desktop, right click and open the 'properties' page. In the 'Target' box, type text in this style (according to your own installations): 'C:\WSJT\wsjtx\bin\wsjtx.exe -rig-name=2mFT8'; The next shortcut could be something like: ' C:\WSJT\

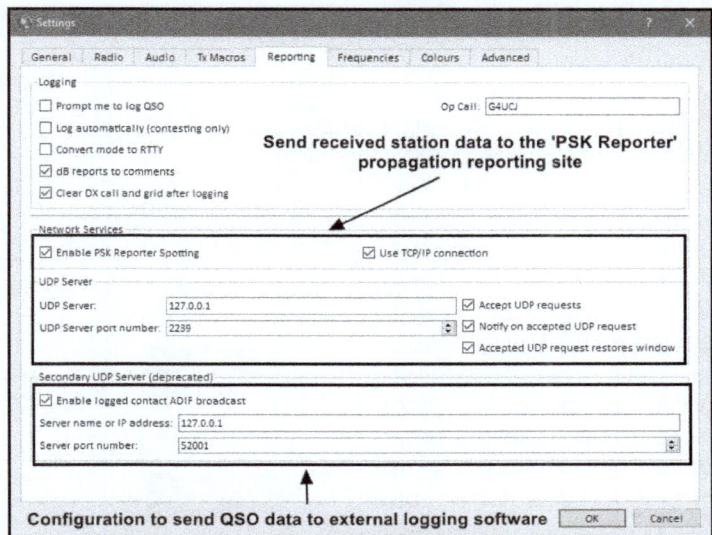

Fig 8.25: WSJT-X Reporting and network configuration.

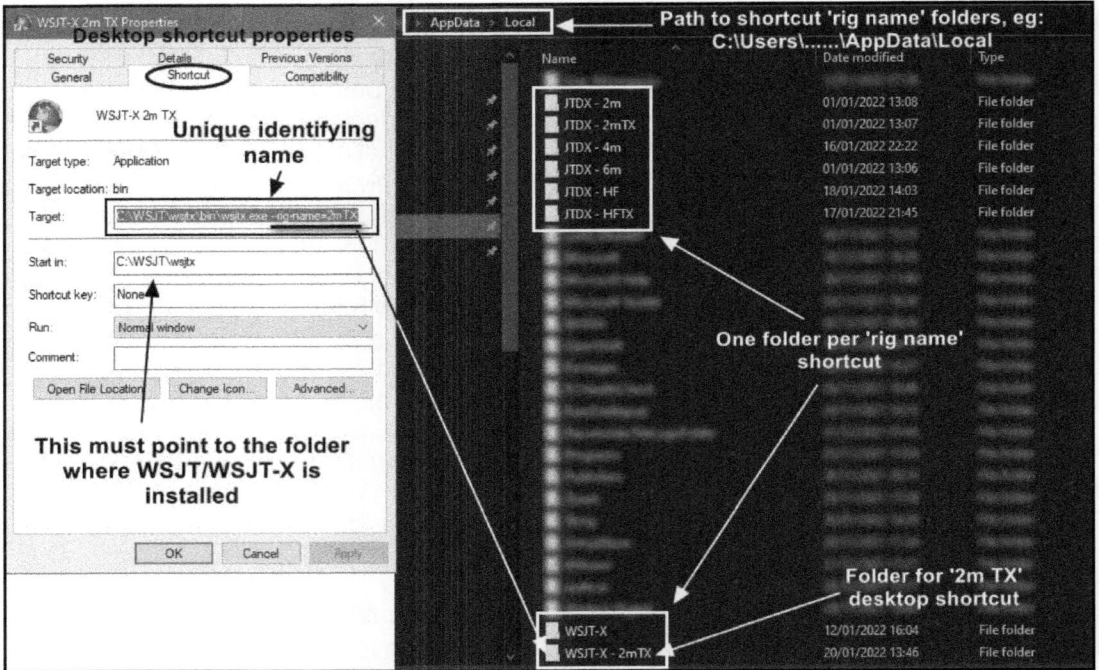

Fig 8.26: WSJT-X/JTDX/MSHV Shortcut configuration.

wsjtx\bin\wsjtx.exe -rig-name=6m'. Each shortcut has a unique 'rig name'. Once this is set up you will be able to run one, two or more instances at the same time. Each shortcut will have a corresponding folder set up in '\Users...AppData\Local'. This folder is set up once the shortcut has been used (see **Fig. 8.26** showing my own version). The configurations (*.ini files), along with logs, master callsign lists, etc., are all saved in the appropriate AppData folders.

WSJT-X also has a waterfall/spectrum display that shows what signals have been received over the past few cycles. The brightness of the signals in the waterfall indicate the relative strength of the received signal and what audio offset is being used. To commence a QSO, double click on a callsign in the main receive window (preferably

one that is calling 'CQ'). The button 'Enable TX' must be pressed before a QSO can take place. The 'Tune' button will transmit a tone that can be used to adjust TX power or the 'ALC' indication on the transceiver (as discussed elsewhere in this chapter). There is a

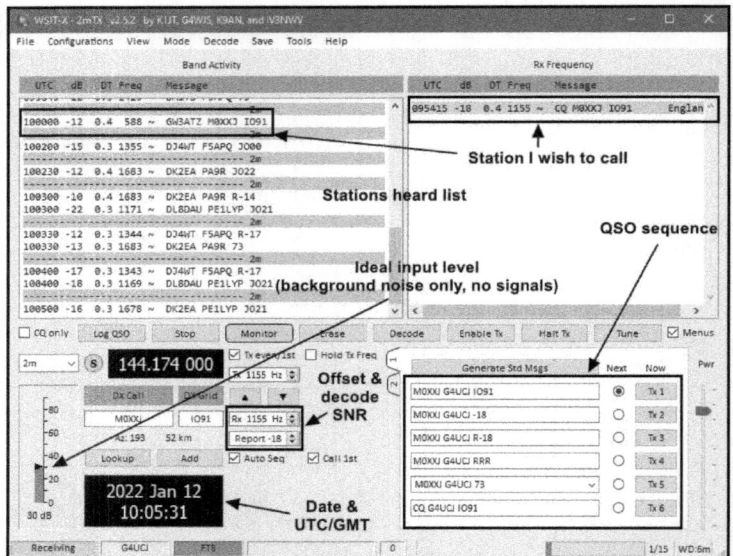

Fig 8.27: WSJT-X main receive window.

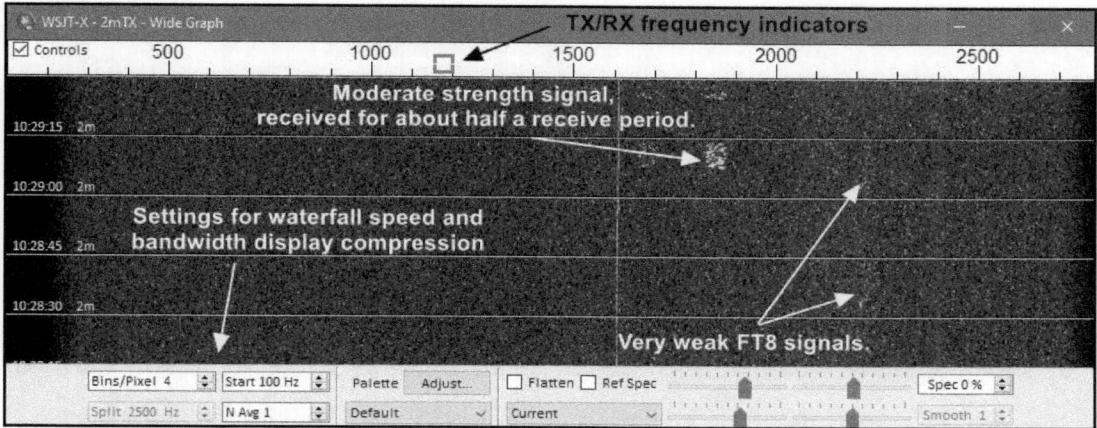

Fig 8.28: WSJT-X waterfall graph.

failsafe. A transmit timer 'watchdog' exists for both 'Tune' and QSO's. The default setting is six minutes. After six minutes of calling with no reply, the software will revert to receive mode. This watchdog timer can be set to the user's preference.

WSJT

Introduction

The original software, the origins of which date back to 2001 and continued in development, up until version 10, which was last updated in November 2015. WSJT 10 offers the following modes: JTMS (for Meteor Scatter or 'MS' as it is commonly known); FSK441, FSK315; ISCAT (A and B); JT6M; JT65 (sub modes A, B, B2, C, and C2) and JT4 (sub modes A-G). These are all modes that are particularly suited to VHF and above, rather than HF due to the propagation types that will be encountered, however some of these modes are occasionally heard on HF. It is worth keeping a copy of WSJT on your computer just in case you encounter one of those modes not covered by the newer WSJT-X.

Installation

WSJT installation is very simple and starts by downloading the latest WSJT Windows executable, which is about 13MB. Double-click the '.exe' file to install it and follow on-screen instructions.

Configuration

To run the software successfully for the first time you may need to run it with Administrator privileges. To do this right-click, choose 'Run as Administrator' and click 'yes' to the prompt that follows. Next, choose 'Options' from the 'Setup' menu item and enter your call sign, grid locator and select the appropriate soundcard and interface connections. Soundcard selection is a little unusual as you need to enter the sound device numbers that are shown on the Console Window, which opens when the programme is started (see **Fig. 8.29**). Adjust the input audio level to show 0dB when only background noise is present.

Operation

After mode and band have been selected and the transceiver has been tuned to the correct frequency for that mode, either of the following methods can be used to instigate a QSO: 1) double click on a received callsign. This will adjust the audio offset to that of the other station. Once the station has finished the previous QSO, or the CQ has ended, press 'TX1' to call the station. If the station responds to your call, progress onto 'TX2', etc., and ending with 'TX5'. To call CQ, the operator will need to press 'TX6'. It is the responsibility of the operator to time the calls to coincide with the beginning of the correct time period due to the lack of rig control in this version of the software. It should be not-

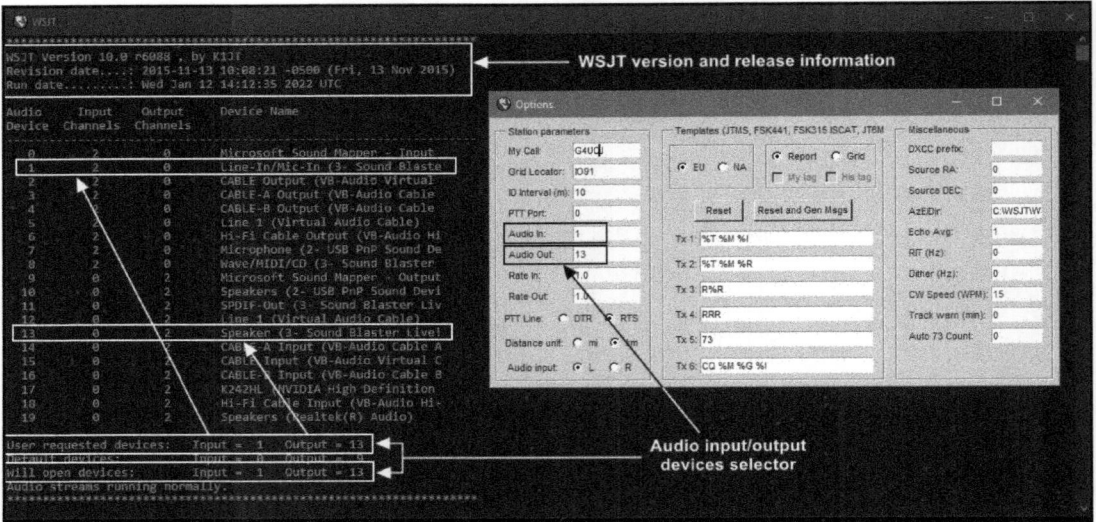

Fig 8.29: WSJT sound device setup.

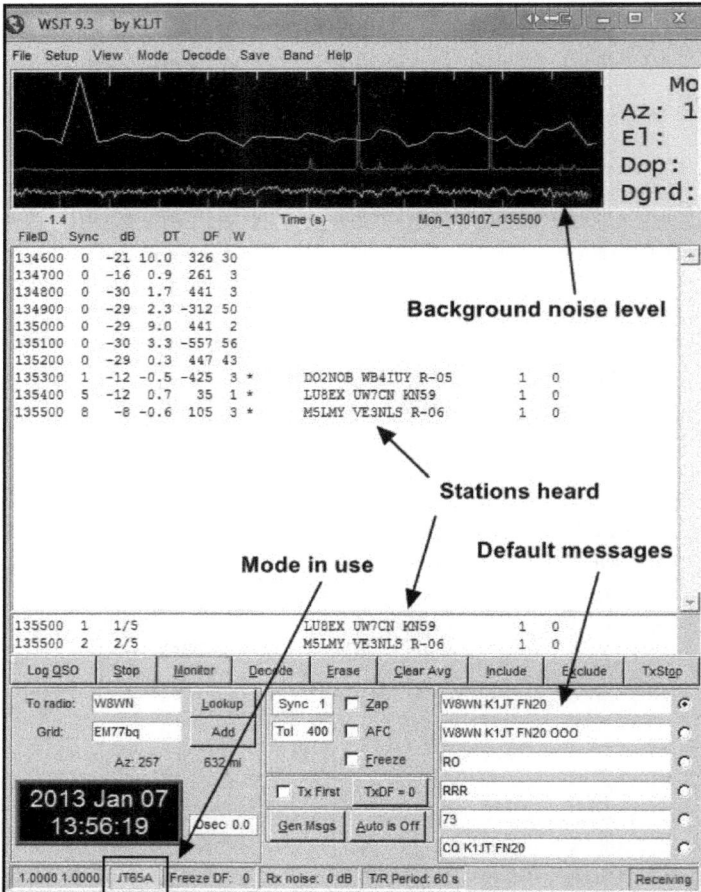

Fig 8.30: Screen layout for WSJT in JT65 mode.

ed that WSJT does not have any rig control except for PTT switching, which is achieved via a serial com port and interface. Additional software that is/was installed as part of WSJT/WSJT-X

MAP65

This software is included as it is a stand-alone programme that installs as part of the main WSJT-X installation. MAP65 is for EME communications using JT65. MAP65 can be used with RF hardware that is capable of providing coherent signal channels for two orthogonal polarisations. The software will automatically match the polarisation of all received JT65 signals within a 90 kHz passband. In single channel mode, MAP65 is extremely effective when dealing with circularly polarised signals. Hardware requirements require a receiver that can convert RF signals into a baseband IQ stream, such as the FUNcube Dongle; RFSpace

SDR-IQ; Softrock range; Microtele-com Perseus and the LinkRF IQ+ range of receivers. Full details are available on K1JT's website.

SimJT

Generates test signals with user defined SNR, allowing experiments to be carried out if no off-air signals are available.

WSPR

Pronounced 'Whisper', WSPR (Weak Signal Propagation Reporter) is a propagation research tool that is de-signed for the sending and receiving of low power transmissions that indicate the presence of a propa-gation path. The results are collated and can be viewed live at wsprnet. org/. Although this version of WSPR can still be used, a far better option would be to download WSJT-X, as WSPR mode is an integral part of the software and has undergone various improvements and updates over the stand alone programme.

JTDX

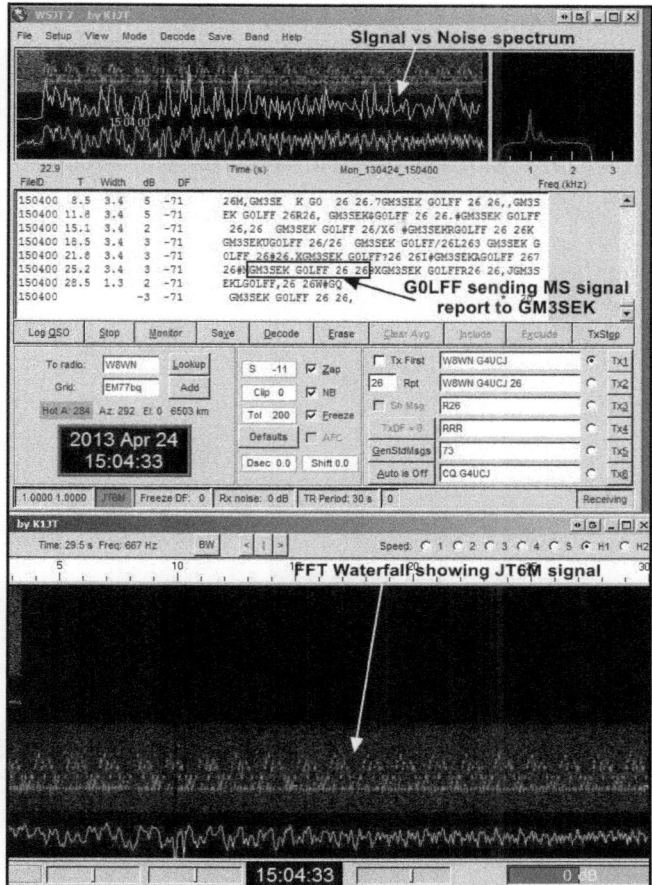

Fig 8.31 WSJT showing JT6M signals.

Introduction

JTDX is a 'fork' (i.e. user modified/custom-ised) version of the well-known WSJT-X soft-ware, adapted by Igor Chernikov, UA3DJY and Arvo Järve, ES1JA (plus contributions from the wider DX community). The rea-son for developing JTDX was to optimise the software for DX use on the HF bands. WSJT-X is primarily aimed at weak signal working in the VHF/UHF/SHF region of the spectrum and, as such, the way the software reacts is tailored for those frequencies and the propagation modes likely to be encoun-tered.

Configuration

The 'settings' menu can be accessed via the main 'File' menu. Once opened, the settings

window brings forth 11 tabs of items to con-figure, it is worth spending some time going through the available options. Building on WSJT-X, JTDX can send your reception data to both the 'PSK Reporter' and 'DX Summit' websites. This is useful as it shows exactly what you have heard/worked over the past 24 hours, but you can also see that data for any other callsign, if they are part of the PSK Reporter network. JTDX, like WSJT-X will send QSO data to your electronic logging software via UDP or TCP. This works very well indeed, providing the port numbers are set up correctly on both programmes. The 'Fre-quencies' tab allows the default frequencies for each mode to be customised according to your preference.

Operation

JTDX accommodates the following modes: FT8/4; JT65/9; T10 and WSPR-2. T10 is a mode specific to JTDX. FT8 has been the most popular data mode for the past couple of years and this shows no sign of slowing down. DXpeditions are now making FT8 one of their chosen modes of operation, ahead of RTTY. The JTDX main window is quite different to the WSJT-X window and has many additional options specifically for the HF operator. On the far right of the main screen is an audio level meter that is marked from 0-90+dB. The level needs to be around 60-75dB for optimum decoding. This differs from WSJT-X which needs a level of just 30dB. To adjust the audio input level, the sound input level needs to be adjusted (right click on the speaker icon in the 'clock' taskbar and select 'Sounds' to access record/playback levels – if you adjust the sounds via the Windows 'Volume Mixer', you will only be able to adjust the output volume to the sound device, rather than the input level).

Fig 8.33 shows the main screen where stations are decoded and QSO's made. There is a busy technical forum on the 'Groups.io' site where issues/ feature requests are discussed. Should a user have a problem they are unable to solve, the forum has many knowledgeable people willing to assist.

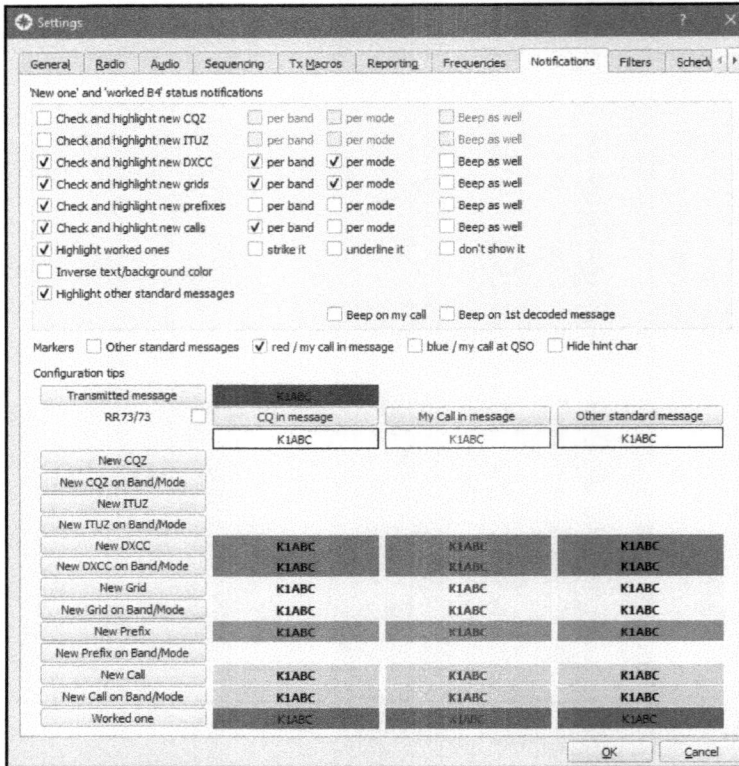

Fig 8.32: JTDX Config window.

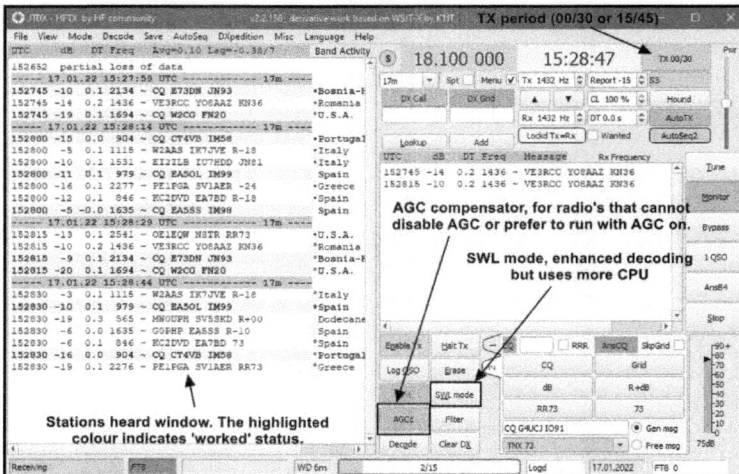

Fig 8.33: JTDX main window.

MSHV

Introduction

MSHV is another variation on the WSJT theme. This programme was developed by Christo Hristov, LZ2HV, together with a group of like-minded amateurs and adds a couple of new modes to the mix, namely MSKMS and PI4 (Pharus Ignis4). This software, like JTDX and WSJT(-X), are free to use and are Open Source. The software/code is licenced under v3 of the GNU General Public Licence (GPL). MSHV is aimed, mainly, at the VHF/UHF weak signal enthusiast, although there is nothing stopping you from using the software on HF.

Installation

Download the latest version from the 'SourceForge' website (both 32 and 64bit are contained in the one installation file). MSHV requires a computer running Windows XP, or newer, in either 32 or 64bit versions. The CPU needs to be a 1GHz or faster, and have 218MB of free RAM. These requirements are easy to fulfil, even with a basic pc. Double click the Installation .exe and installation begins. The licence agreement is shown on the next screen, you have to agree with the licence in order to proceed with the installation. Failure to agree with the licence will terminate the installation.

Configuration

Work your way through the menus, where you can configure sound devices; rig/transverter control; network and interface; decoding options; etc. Should the user have a relatively fast/modern computer, the 'decode' menu settings can be set to the maximum for each mode without fear of the PC lagging behind.

Operation

Select the mode and band you wish to operate on. Click the 'monitor' button to commence re-

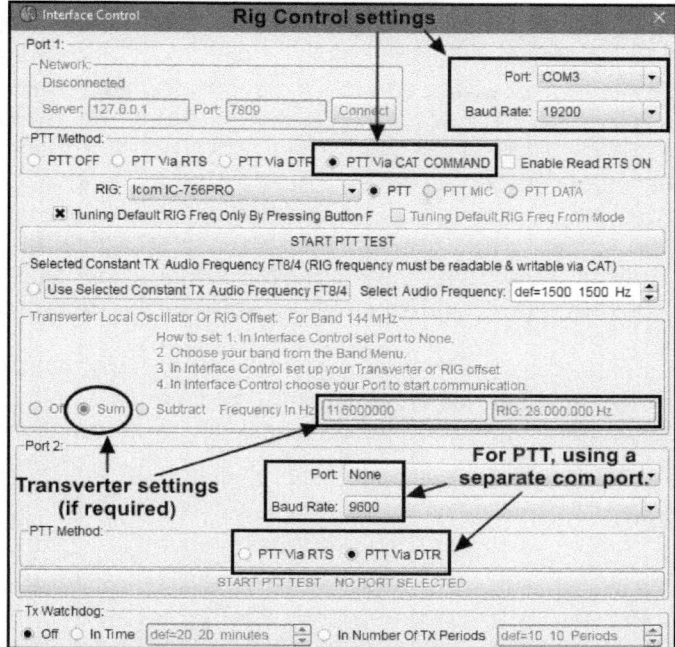

Fig 8.34: MSHV rig control configuration.

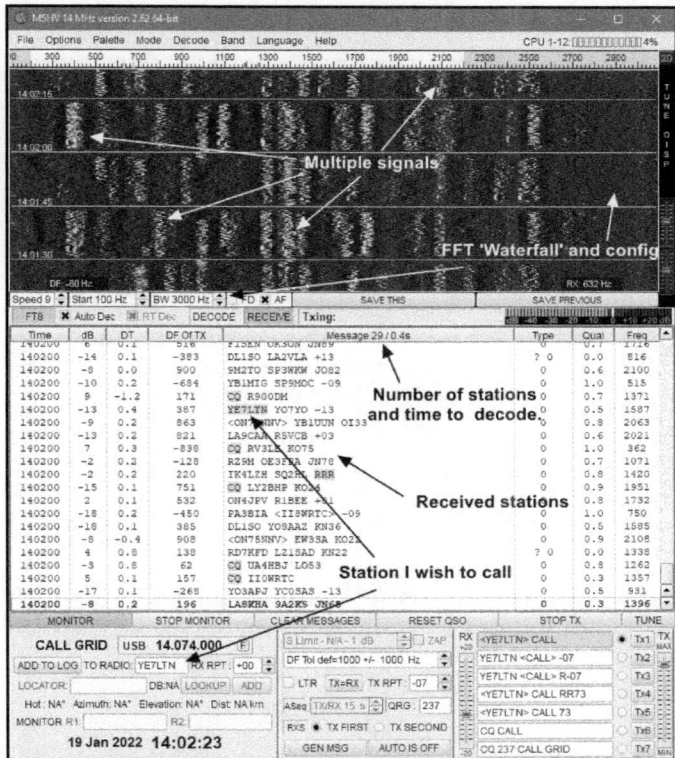

Fig 8.35: MSHV Main RX/TX window

ception of the chosen mode. When you wish to call a station, double click the callsign in the main window. The callsign is placed into the QSO/macro box on the right hand side of the window. To call the station, press 'TX1' at the start of the correct transmit period. Upon receipt of the appropriate response from the other station, work through the TX macros or select 'Auto' to let the software do it for you. That may sound lazy but the software will ensure the correct reposes are sent and will send them at the correct time. Left to the human operator, either, or both, of these parameters could be actioned incorrectly or out of sequence.

ROS

Introduction

ROS was devised and coded by the Spanish radio amateur Jose Alberto Nieto Ros, EA5HVK. Controversial when first introduced, due to questions regarding the legality of the mode for use in amateur radio, particularly within North America. ROS uses Multiple Frequency Shift Keying (MFSK) along with sophisticated encoding techniques to provide a very robust data mode that's ideally suited for use under poor conditions. ROS has never gained the popularity of some other modes but it can still be heard regularly.

Installation

Download and extract the zip file to a location of your choice. Once the file has been extracted, double-click on the Install.exe file to complete the installation process. The latest version (v7.40, last updated in 2016) has an improved installation process and now puts shortcuts and programme menu entries in automatically. I did experience an unusual error when installing, whereby ROS would not install due to there being a missing DLL. This DLL is presumably included as part of the setup executable, as it appeared in the main ROS installation folder. The 'missing' DLL (PSKReporter_1_6.dll) was then copied into the Windows 'System32' folder, as required by ROS and the PC restarted. This time installation went smoothly. A very odd error and made me wonder if the software was truly compatible with Windows 10. It appears that it does, indeed, run perfectly under Windows 10. Finding ROS signals to decode though proved to be another thing entirely!

Configuring

The first step is to open the 'Configure' menu item and choose 'Operator' to enter your station information. This is followed by the rig setup, where you can choose the type of

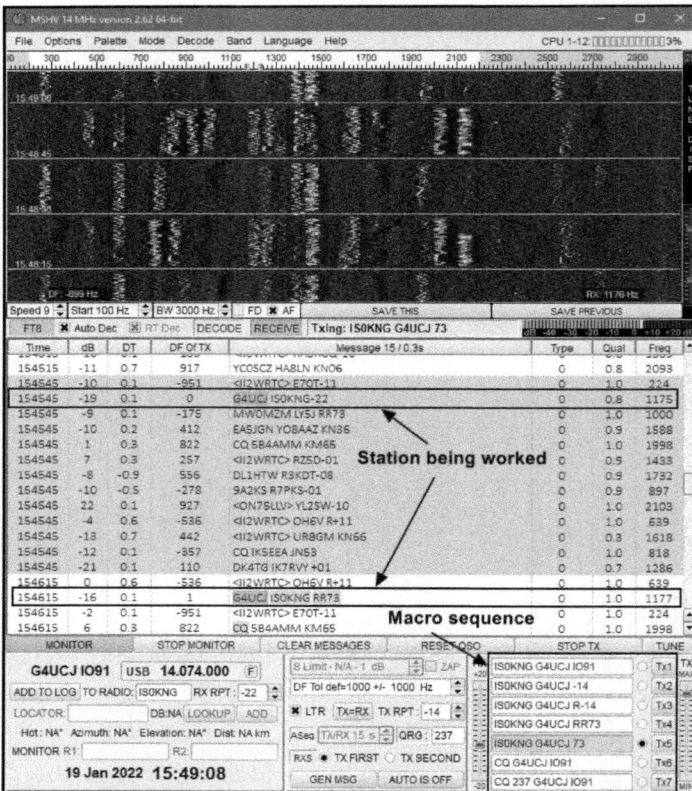

Fig 8.36: MSHV in QSO.

rig control you want to use and select your rig. ROS makes this process very easy as it automatically scans the COM ports to locate your rig, then configures the data rate, etc. The final step is to select the appropriate sound device for input and output. In the

central area, just above the type-ahead buffer, you will find the macro buttons, which are pre-configured with a selection of useful texts. These can be edited by right-clicking over any of the buttons.

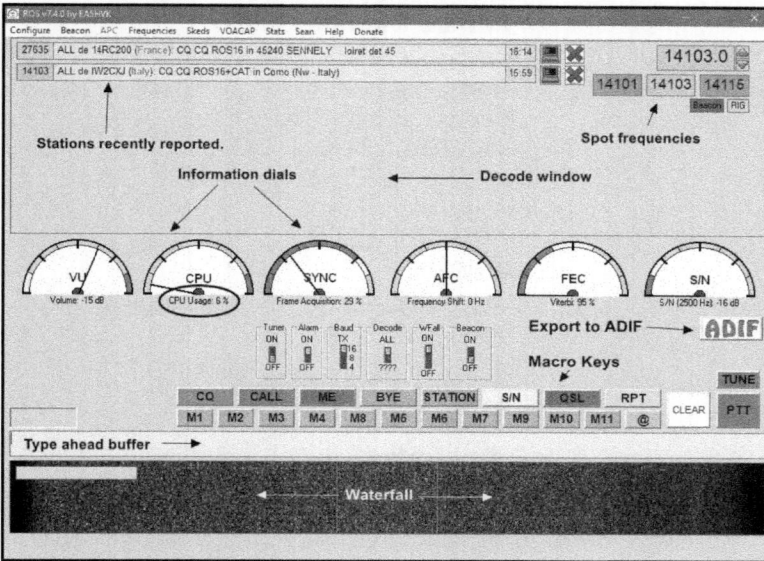

Fig 8.37: ROS main screen.

Software:	FLDIGI	HRD/DM780	MixW 3
Website:	www.w1hkj.com	www.hrd.com	www.mixw.net
Trial Period:	—	30 days	15 days
Current cost:	Free	Licence (with 1 year support and updates): $99/€87/£73	Full version: £45, $61, €53 Update from v2: £15/$18/€17.50
Software:	MixW 4	MultiPSK	TrueTTY/SeaTTY
Website:	www.rigexpert. com/software /mixw-4	www.f6cte.com	www.dxsoft.com
Trial Period:	15 day	—	30 days
Current cost:	Full version £45, $61, €53 Update from v2/v3: £15/$18/€17.50	Free* Full version $40/€35/£29	TrueTTY $39/€35/£29 SeaTTY $49/€39/£36

* Unregistered copy limited to 5 minutes of 'professional' modes decoding. All 'amateur' modes are unrestricted

Table 1: Multimode data decoding software comparison table.

Operation

During idle, waiting for a signal to appear, ROS seems quite processor hungry. Using ROS with a modern 6 core Intel i5 processor, the CPU usage according to ROS is 6%. For comparison, SDR Console v3 which is running the Airspy HF+ Discovery, is using 0.2% CPU

The author has made ROS extremely easy to operate, especially if you are able to link ROS with your rig. At the top right of the screen you will see the band and channel information. You can use these controls to select your operating frequency. To the left of that panel you will find the activity reporter that uses your Internet link to flag-up other ROS activity across the bands. The central area also contains a modem dashboard of indicators that may at first look a little daunting.

These are simply gauges that show the quality of the received signal – the more greens you get, the better the signal. You will also note that ROS provides a measured signal-to-noise ratio. It's customary to use this as the signal report, rather than the subjective RST system (this is known as 'RSQ' – Readability, Strength, Quality)'. It is also important to note that ROS is a very robust system with built-in error correction, so you shouldn't need to repeat key information such as name, QTH etc.

Resources

Super Anti Spyware:
www.superantispyware.com

Malwarebytes:
www.malwarebytes.com/mwb-download

Oracle VirtualBox:
www.virtualbox.org,

VMware 'VMware Player' :
www.vmware.com/products/player

Fldigi:
www.w1hkj.com/files/Fldigi

Ham Radio Deluxe: DM780
www.hamradiodeluxe.com/downloads

MixW v3:
http://www.mixw.net

MixW v4:
www.rigexpert.com/software/mixw-4

MultiPSK:
f6cte.free.fr/index_anglais.htm

TrueTTY and SeaTTY:
www.dxsoft.com

WinWarbler:
www.dxlabsuite.com/winwarbler

WSJT-X:
physics.princeton.edu/pulsar/K1JT/wsjtx.html

WSJT:
physics.princeton.edu/pulsar/K1JT/wsjt.html

JTDX:
jtdx.tech/en

MSHV:
sourceforge.net/projects/mshv/

ROS:
rosmodem.wordpress.com

9
SSTV

by Leo Ponton M0NNQ

SSTV or Slow Scan Television was developed in 1957 by Copthorne Macdonald. While broadcast television transmitted 25-30 frames per second using 8MHz bandwidth providing smooth motion imagery, SSTV transmitted one frame over a period of 8 to 120 seconds and only used 6 kilohertz bandwidth – smaller by a factor of over a thousand.

Both the Russian and American space programs used the technology to send images from space After SSTV was legalised for amateur use in the USA in 1968 it became used predominantly by amateur operators to send and receive still images.

Equipment was expensive, complex and often home-built. A television camera utilising a type of cathode ray tube called a vidicon was required to capture the image and some sort of persistent monitor was needed to display the image such as a radar display cathode ray tube. Circuitry was often constructed with vacuum tubes (known as valves in Britain) and early transistors as well

Fig 9.1: CQ SSTV.

as surplus and repurposed devices.

Broadcast TV in the UK became standardised on a picture made of frames of 625 horizontal lines . To achieve smooth motion the frames were broadcast at a rate of 25 frames per second . The frame rate is time-based on the frequency of AC mains – 50Hz in the UK . Each frame comprised two scans with the lines interlaced, so 50 (or 60) scans per second were transmitted to provide the 25 (30) frames. The frames were scanned from top to bottom and the time taken for one scan would be 0.02 (0.017) seconds.

The name Slow Scan TV, derives from the much longer time taken to scan from top to bottom of the picture. Because the transmitted data is stretched out over time, the bandwidth is much narrower covering a few kHz rather than upwards of 8MHz. Think of it like a coiled telephone wire – as you stretch it out the width reduces. The same amount of wire, but spread over a greater distance.

Fast Forward to the Present Day

While the era of the vacuum tube and, to a great extent, transistors has passed by, SSTV remains as an analogue mode albeit processed digitally. Equipment requirements are more straightforward, utilising what we already have – a computer and a transceiver. Digital SSTV, although seemingly a development of traditional analogue SSTV is essentially a process of file transfer incorporating error correction and does not truly follow the line-by-line horizontal scanned paradigm of analogue TV and SSTV.

Images are now captured using CCD image sensors in either a camera or scanner. Any image can be used and does not have to be captured directly - images from the internet could be used, for example. We have also moved on from the need for a persistent monitor as images are received into computer software, decoded, displayed and can be saved for later viewing. Links to some of the software and websites mentioned can be found in a reference list at the end of this chapter.

Easy Listening

As you have probably already configured your computer and radio to talk to each other I will concentrate on how to set up the software that is required for SSTV. you only need a computer and an internet connection.

It's easy to make a start with slow scan television even if you don't own a shortwave radio - we you can use one of the many SDR websites combined with one of the applications for decoding SSTV. This also means that if you have poor reception in your location, you'll still be able to receive the signal from anywhere else in the world and download SSTV pictures. In fact, using this technique it's possible to download pictures from the International Space Station (ISS) even if it isn't passing overhead. All you need to do is select a web SDR that is close to its path.

I'll demonstrate the process of accessing WebSDR for SSTV using macOS as the Mac configuration is possibly the most difficult way to do it due to some considerations with the Mac relating to permissions and signed images.

First of all, a little about WebSDR and SDR (software defined radio). SDR is a technology that replaces most of the traditional hardware of a radio with computer programs and the hardware they run on. The analogue signal is captured by a discrete RF front end and is converted to digital information using and analogue-to-digital converter. From this point onwards, all signal processing is performed by the computer programs

until the output is converted to audio via a digital-to-analogue converter.

Web-based SDR uses a software application that enables the signals received by an SDR to be made available on a website. WebSDR is the most popular and uses server software written by Pieter-Tjerk de Boer (PA3FWM). The technology permits many users to access one SDR simultaneously on the same or different frequencies from anywhere in the world. WebSDRs are very individual in their configuration and location, using a variety of antenna types and band coverage. Another similar system called KiwiSDR uses a BeagleBone embedded computer and is limited to four simultaneous connections but it does have a well-designed and standard interface.

For our purposes, indeed for most purposes, we can use either. There are maintained online lists of available SDRs. I have provided a couple of suggestions in the reference section. These lists indicate current status, location, antenna type, number of users online and more. As most SSTV activity occurs on the 20m band we can usefully chase the daylight around the earth and, hopefully, receive SSTV at any time of day!

Now we need to know a little about the SSTV software that we're going to use:

Linux-based computers have a package called QSSTV. QSSTV was written by Johan Maes (ON4QZ) and is actively maintained. It is easy to install using the Linux distro's package manager and interfaces well with Hamlib and Rigctl, the libraries and program that provide CAT control of the transceiver from the computer.

Windows users have a choice of MMSSTV and EasyPAL. Sadly the author of EasyPal, Erik Sundstrup (VK4AES), died in 2015. MMSSTV by Makoto Mori (JE3HHT) was last updated with the 1.13a release in 2010 but has recently been brought up to date by Eugenio Fernández (EA1ADA) and it is now called YONIC. YONIC uses Omni-Rig to interface with the transceiver.

Macs have MultiScan 3B by Sergi Loudanov (KD6CJI). The latest version is 1.9.5 (2010) but is still fully functional on the latest version of macOS at the time of writing, Monterey.

Installation on Mac

Please read this even if you don't use a Mac as the general principals are the same for all platforms. Download the latest version of MultiScan 3B.

Open Finder and navigate to Downloads, then double click 'multiscan 3b-v1.9.5.zip' or you can open it from the browser (your zip file name may be slightly different). This will extract the zip file into Downloads – there is only one file within and that is the application.

Now it gets a little tricky. Double click the application - it's the one called 'MultiScan 3B' with an icon that looks like an old television. You will probably receive the following message:

"MultiScan 3B" cannot be opened because the developer cannot be verified.

Don't panic! This is your Mac protecting you from malicious software by being cautious. Click on the Apple logo at the left end of the top bar (the one with the date & time at the other end) then choose System Preferences…, click Security & Privacy and make sure General is selected. Unlock the panel if you need to by clicking on the padlock at the bottom left and entering your password.

On the Security and Privacy dialog you should see "MultiScan 3B was blocked from use because it is not from an identified developer. And a button next to that that says 'Open Anyway'. Click the button.

On the next warning dialog click the 'Open' button. The application should open up full screen.

Move the file you opened to a more sensible location – I put it on the Desktop. You can just drag it onto Desktop in Finder.

You now need to install an application, Soundflower, a virtual internal audio device that will allow you to pass sound between applications. There are alternatives such as Loopback, but we'll stick with Soundflower here. Note that this is only required for using the web-based SDR. Connecting to a transceiver will use the standard audio connections.

For Linux try PulseAudio Volume Control and in Windows VB Cable is easy to use. Download the latest versionfrom Github.

Installing Soundflower may need similar steps to installing MultiScan 3B regarding Security and Privacy. You will need to double click the downloaded Soundflower.dmg file. This will open to a folder containing a Soundflower.pkg file. Again, double click the pkg file (you may need to hold down the ctrl key) and click the continue then install button on the dialog and enter your password if requested.

You should now have the SSTV application MultiScan 3B installed as well as a virtual soundcard.

Open Multi Scan 3B and access the preferences below MultiScan 3B on the top bar. Set Audio Input and Audio Output to Soundflower (2ch). Close the dialog by clicking on the red X.

In (Apple logo) / System Preferences / Sound set the output and input to Soundflower (2ch). Open your browser and navigate to a SDR site. A link to one is provided in the reference list.

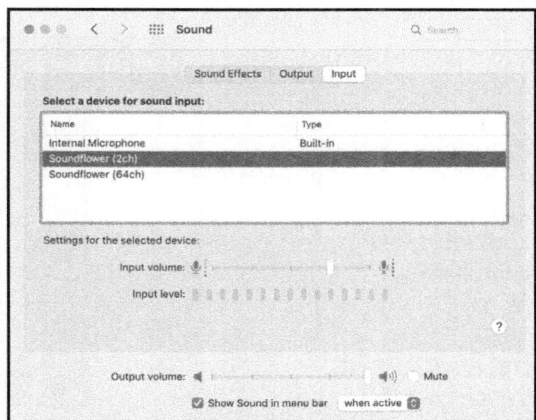

Fig 9.2: Selecting the 2ch input and output in Soundflower.

A bit about sound on Chrome
(and probably other browsers in future):

As browsers move towards better security for users, i.e. HTTPS rather than HTTP, Google have disabled sound on untrusted websites. HTTPS requires a certificate and this adds cost to hosting the website.

At the present time, if you have no alternative to Chrome, this can be worked around by allowing the sound on a site-by-site basis. To access the "Site settings" you should click, in the address bar, on the text that says "Not secure". Find the sound setting and set to "Allow". Chrome will remember this for future visits to the same website. Sometimes there will be a "Chrome audio start" button which performs the same task.

This is explained in detail by Tony G1HMO on the Hack Green WebSDR website Hack Green, incidentally, is a HF WebSDR located at the Nantwich Secret Nuclear Bunker, formerly R.A.F. Hack Green, now a working museum.

Installation in Ubuntu Linux

You can use the Ubuntu Software manager to install QSSTV and PulseAudio Volume Control or open a terminal (CTRL+ALT+T) and type the following:

sudo apt update
sudo apt install qsstv pavucontrol

Configuration and use is intuitive and straightforward:

Installation in Windows 10

Download the latest Yonic build from Hamsoft. The file

is a zip file containing the installer MMSSTV_YONIC.exe Unzip this file and double click to install. The installation may proceed in

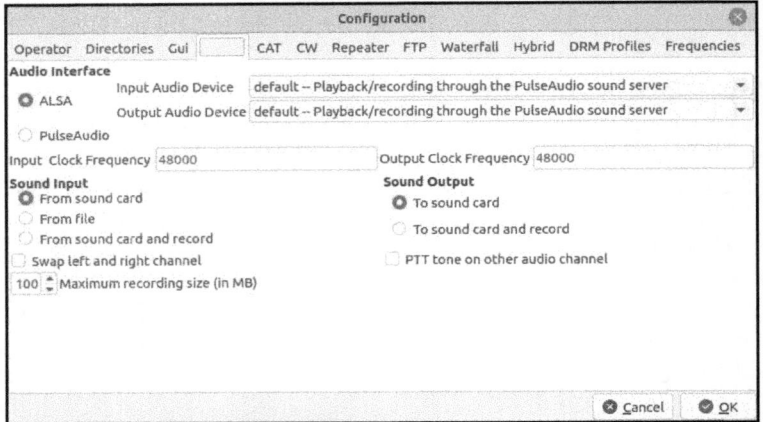

Fig 9.3: QSSTV basic configuration.

Fig 9.4: Pulse Audio configure playback.

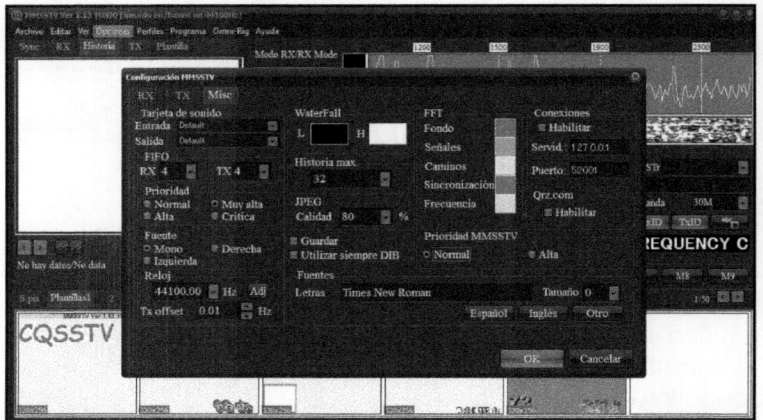

Fig 9.5: YONIC configuration.

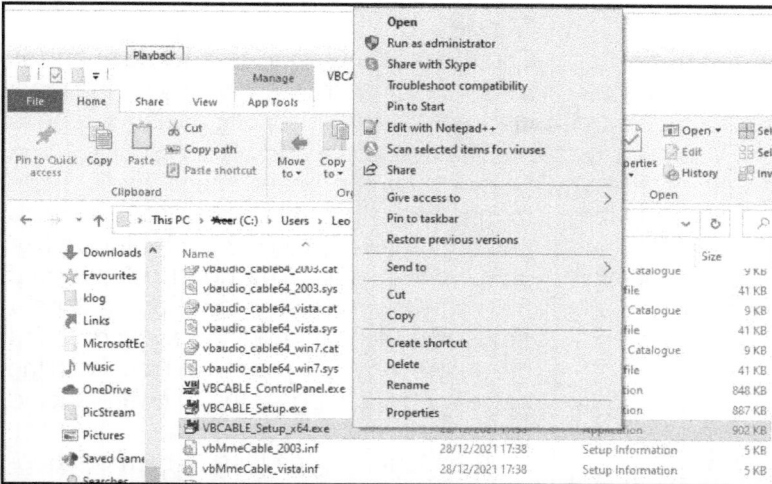

Fig 9.6: Run VBCABLE_setup_x64.exe as administrator.

Spanish – just accept everything. If it has installed in Spanish it's easy to change this in 'Opciones', selecting Inglés from the bottom right of the dialog.

From Vb-audio download the VBCA-BLE_Driver_Pack43.zip. This zip file contains installers for 32 bit and 64 bit Windows versions from WinXP –to Win10.

You will need to extract the zip then, as administrator, run either the 32 bit or the 64 bit installer. To run as administrator, right click on the exe file and select 'Run as administrator'. Once the installation is complete, you will need to reboot. The virtual cable will then be available as a sound device to select in the options for YONIC.

Testing

Choose a site from WebSDR ensuring that it receives on 14.130 MHz. Set the audio output to internal speakers or whatever you normally use (you don't want to use Soundflower yet). Enter 14230 into the frequency field – it's normally in kilohertz. Select 'USB' and 'Wider'.

Hopefully you will now be able to hear SSTV, it

sounds like an old fax modem. If you don't hear anything and you have checked the audio output is set to your normal device, e.g. speakers, then look for a button that says 'Audio Start' or 'Chrome Audio Start' if you're using Chrome. Check that you allowed audio for the website as described earlier – it's easy to forget if you're switching between sites. Once you can hear SSTV, switch the audio input and output to 'Soundflower 2ch'. If all you hear is noise try a different web SDR somewhere that you would expect to be in daylight.

If all is well you should now start to see images appearing in the MultiScan 3b window.

Workflow for Linux and Windows is similar but you should only need to select the virtual sound device in the SSTV application, not at system level.

Note that for on-air use of the SSTV applications on all platforms, you will not need to use the virtual sound device but, instead, the device that is configured for communication with your radio.

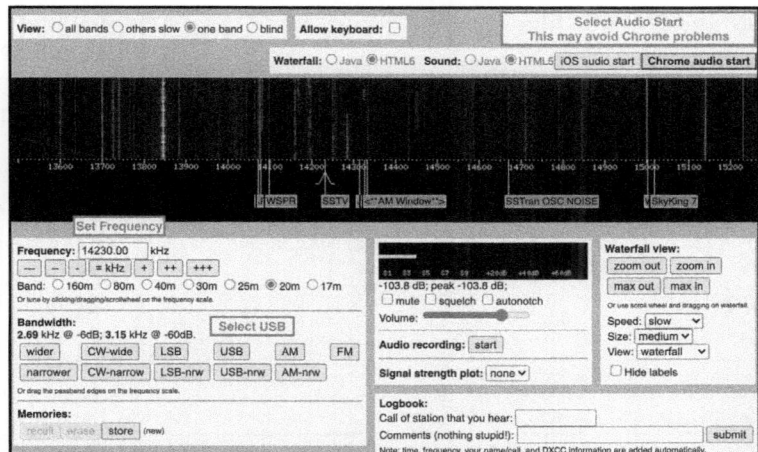

Fig 9.7: Setting the options in the WebSDR.

Interfacing

Interfacing the computer to the radio has been covered in this book. The same interface applies for SSTV, however, for full control of the transceiver from the SSTV application an interface is required with CAT control as well as an audio interface. You may find that you will need to adjust the send and receive volumes. This will depend on the application you are using and your transceiver.

Where's the Action?

Like all other modes in amateur radio SSTV tends to concentrate around certain frequencies. Suggested frequencies are listed in table 9.1. It's not a comprehensive list and is based on the RSGB band plans. Sometimes there simply isn't any traffic and the bands are quiet or closed, but the busiest area is likely to be on the 20m band at 14.230 MHz, in Europe this may also be at 14.233 MHz. Because of the nature of propagation on this band activity at night time may be greatly reduced or non-existent. If you only wish to receive images then in this instance you can use a web-based SDR located in a country in the daylight area.

You may also find that users of other modes intrude on the SSTV allocated frequencies, particularly during competitions. When you consider the relatively light use of the allocation as well as the weak signals that may not be heard, such transgressions are understandable.

Band	Frequency
70 centimetres	432.500 – 432.625MHz
2 metres	144.500-144.600MHz
6 metres	50.500-52.000MHz
10 metres	28.680 MHz
15 metres	21.340 MHz
20 metres	14.230 MHz, 14.233 MHz
30 metres	10.132 MHz – Use narrow mode MP73N
40 metres	7.040 MHz, 7.043 MHz, 7.165 MHz
80 metres	3.735 MHz

Table 9.1. Suggested frequencies

Modes

There are many modes and sub-modes available to use in SSTV, but by far the most common and most popular analogue modes are Robot, Scottie and Martin. Scottie 1 and Martin 1 are the better image quality sub-modes and take a little under two minutes to send or receive. Scottie 2 and Martin 2 take about one minute. The good news is that the software you use should automatically determine the mode being received. If you are answering a CQ you should reply with the same mode, but again the software should take care of this for you.

The Martin modes tend to be more popular in Europe and the Scottie modes pretty much everywhere else. In practice, it makes no difference particularly if you are working DX and the QSO will continue in whichever mode it was established.

Once you have some experience sending and receiving, experiment with other modes such as SC2 for 80m. On the WARC bands you should use MP73-N, MP110-N and MP140-N narrow modes as the bands themselves are very narrow.

Transmitting

SSTV is a 100% duty cycle mode. This means that the RF power amplifier is running at the selected power for the whole duration of the transmission. Most modern radios are not designed to be run at full power continuously and you should check the rating in the user manual for the transceiver. Sensible precautions are to keep an eye on the ALC to ensure you aren't overdriving and also to keep the output power below 50%, i.e. less than 50 watts on most radios. Of course, the vents, fans and heatsinks will already be clear of obstructions! These precautions apply equally to the power supply.

Prepare

Before you start it is wise to prepare some images with overlaid templates. A short QSO typically follows the pattern:

- CQ
- Reply with report
- Report with 73
- 73

and this means that you will need to modify the content of at least two images – the CQ doesn't carry any variable information.

Fig 9.8: A QSO in progress. The template on the left is filled from the information entered in the bottom panel.

The application you use, whether MultiScan 3B, QSSTV or MMSSTV/YONIC, will have a template or overlay system to assist with this. Simply put, you provide an image and create an overlay template. As the QSO progresses you enter information into a form, e.g. callsign & report, and the overlay system populates the template with this information. The image and the template together are then transmitted as a single image. There is some flexibility in that you can use the same template with a different image and vice versa and, of course, you can have several templates defined. You should make sure that the text positions in the template overlay the image in a way that they remain readable.

Images should be clear without too much fine detail. Simple stills from cartoons, photos of your local area even of yourself are all worth trying, but it's probably best to avoid overtly political images or images that some might find embarrassing or offensive. It's also a good idea to have several images prepared in advance to the right size.

The report in SSTV is similar to both SSB and CW but differs in the third element which reports video quality:

Readability

1 Unreadable
2 Barely readable, occasional words distinguishable
3 Readable with considerable difficulty
4 Readable with practically no difficulty
5 Perfectly readable

Strength

1 Faint—signals barely perceptible
2 Very weak signals
3 Weak signals
4 Fair signals
5 Fairly good signals
6 Good signals
7 Moderately strong signals
8 Strong signals
9 Extremely strong signals

Video

1 Barely perceptible
2 Poor
3 Fair
4 Good
5 Excellent

ISS: Pictures from Space

The Amateur Radio on the International Space Station (ARISS) program aims to inspire an interest in science, technology, engineering and mathematics among young people.

Periodically, throughout the year, the Russian ARISS Team transmit SSTV from the Russian Service Module of the International Space Station (ISS) using a Kenwood D710 transceiver. The FM transmission is on 145.800 MHz at about 25 watts. The mode used is PD120 and it is possible to receive 12 images in a single overhead pass.

The best way to receive the images (this

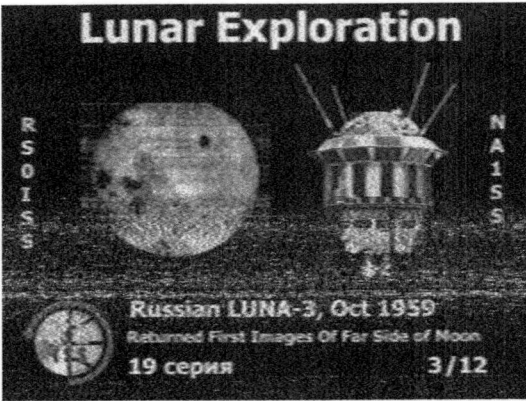

Fig 9.9: ISS transmission 29 Dec 2021, captured from a Web SDR in Johannesburg.

is a one direction communication) is to use a web-based SDR as described earlier. It's straightforward enough to receive on a handheld 2m using a ¼ wave antenna, but in IARU Region 1 (British Isles, Europe, Africa) you should select WFM as the transmission uses 5 kHz deviation – IARU Region 1 uses 2.5 kHz by default.

Transmission schedules are posted on the ARISS SSTV blog.

There are online calculators such as Westweather to determine when the ISS is passing overhead and if there isn't a convenient pass you can use a web-based SDR under its path anywhere in the world.

The European Space Agency (ESA) have published a series of video tutorials to help you to set up your computer to receive these transmissions. They cover the three main platforms MacOS, Ubuntu Linux and Windows 10 and although the operating system versions are superseded, the principals still apply. These videos are a great help for setting up for normal SSTV operation as well but the Windows tutorial uses RX-SSTV (receive-only) not MMSSTV/YONIC.

Finally...

As with all branches of amateur radio, there are many variations and possibilities and the potential to combine with other activities. SOTA or POTA SSTV, QRP SSTV, contests

and so on.

This has been presented as an introduction to the absolute basics of this fascinating discipline and, unlike the early days, there is now a wealth of information available simply by searching the internet. I have included some links to further information in the reference section, but these things do change and may be replaced or removed. If that has happened I apologise, but the benefit is that they have more than likely been superseded by more up to date information. Just search the internet with "SSTV" in the query.

References

Online list of available SDRs: http://www.websdr.org/ and http://kiwisdr.com/public/
MultiScan 3B from https://www.qsl.net/kd6cji/
SDR site such as http://hackgreensdr.org:8901/
Hack Green WebSDR: http://hackgreensdr.org:8901/Chrome.pdf.
Yonic build from https://hamsoft.ca/pages/mmsstv-yoniq.php
Vb-audio: https://vb-audio.com/Cable/ download the VBCABLE_Driver_Pack43.zip
http://www.websdr.org/ (for testing)
Transmission schedules: https://ariss-sstv.blogspot.com/
Online calculator: https://www.westweather.co.uk/isslive
European Space Agency (ESA) video tutorials:https://www.esa.int/esearch?q=sstv
Further information links: The Slow Scan Companion:
https://batc.org.uk/wp-content/uploads/SlowScanCompanion.pdf
ARRL article from QST Magazine: http://www.arrl.org/files/file/Technology/tis/info/pdf/99753.pdf
Revisiting Slow-Scan Television - A Second Look At First-Generation SSTV
By John Magliacane, KD2BD
https://www.qsl.net/kd2bd/sstv.html
Soundflower: https://github.com/mattingalls/Soundflower/releases/download/2.0b2.dmg

10
Remote Operation of Amateur Stations

by John Regnaut, G4SWX

Introduction

In recent years advances in computing hardware and cheaper Internet Protocol (IP) networking have enabled many to operate an amateur radio station by remote control. This chapter will cover the UK licence conditions and a range of technological and operational issues that remotely controlling an amateur station entail. At the end of the chapter, you will find a list of references for some of the matters discussed in this chapter.

Remote control of all aspects of an amateur radio station opens a very wide range of possibilities from individuals with access to a remote location where they can establish an amateur radio station to a radio club enabling its members, who do not have any possibility of erecting outside antennas, to get on the air. Of increasing relevance is the ability to enjoy amateur radio from a remote location with lower noise and greater possibility of planning permission for antennas than is the case in most built-up areas.

UK Licence Conditions

Some countries do not permit remote operation of an amateur radio station but in the UK, we are fortunate that since 2006, amateur licences have specific clauses permitting remote operation of an amateur station. The author considers that these particular amateur licence clauses, which might not be familiar to many radio amateurs, do set boundaries on the technology that can used in remote amateur stations. The URL to the Ofcom PDF can be found in the references section at the end of this chapter.

The first specific term is easy, remote operation of a UK amateur radio station is permitted

10(2) Subject to Clause 10(3), the licencee may also conduct Remote Control Operation of Radio Equipment (including, for the avoidance of doubt, Beacons) provided that any such operation is consistent with the terms of this Licence.

The main constraint to this is laid out in Clause 10(3):

10(3) This Clause 10 does not permit the licencee to install Radio Equipment capable of Remote-Control Operation for general unsupervised use by other Amateurs.

This is fairly obvious; in the physical world you are not allowed to let unsupervised others operate your amateur station. The same applies to remote control. However, remote operation of a club amateur station by the authorised members of that club is permitted. A fully featured remote amateur club station is technically quite difficult to achieve but could provide a route for some to get on the air from a club station when their personal circumstances would not normally allow operation of an amateur radio station.

Clause 10(2) specifically instructs that any remote operation must be consistent with the other terms of the amateur licence. Of particular note is that pay-to-operate remote amateur stations that are appearing in other parts of the world are not permitted in the UK.

1(1) The licensee shall ensure that the Radio Equipment is only used:

(b) as a leisure activity and not for commercial purposes of any kind.

There are two other clauses within the amateur terms that are of relevance:

10(4) Any communication links used to control the Radio Equipment or to carry Messages to or from the Radio Equipment in accordance with Clause 10(2) must be adequately secure so as to ensure compliance with Clause 3 of this Licence. Any security measures must be consistent with Clause 11(2) of this Licence.

Clause 3 is concerned with who can operate an amateur radio station in the UK so Clause 10(4) is about ensuring that any remote-control solution will only allow those authorised to access the amateur station. In some ways this clause is complimentary to Clause 10(3) but it is there to ensure that amateurs do ensure that security measures are incorporated in any remote-control solution. The term 'adequately secure' can be interpreted in many ways but the best idea is to look at a real world parallel. If your amateur station is located in an outside building do you have locks on the door and windows to prevent others gaining entry and possibly an alarm system. Similarly, with a remote-control solution do you have good access controls on electronic forms of access and can monitor when the station is on the air.

The words: "Any security measures must be consistent with Clause 11(2) of this Licence seems to leave some amateurs confused although the explanation is quite simple. Clause 11(2) states:

11(2) Unless the Radio Equipment is being used for the purposes of clauses 1(2) or 1(3) in the UK: (b) Messages sent from the station shall not be encrypted for the purposes of rendering the Message unintelligible to other radio spectrum users

In the respect of remote operation of an amateur station this clause specifically outlaws the use of encrypted links operating under the terms of an amateur licence (see later in the chapter for more detail). It does not prohibit the use of encryption over the Internet or private LANs. In that respect encrypted Wi-Fi connections, made using licence exempt equipment, at both the control and radio end are permitted.

10(5) The use of any such communication links referred to in Clause 10(4) must be fail-safe such that any failure will not result in unintended transmissions or any transmissions of a type not permitted by this Licence.

If your communications link fails, for whatever reason, you must ensure that the remote station is not left transmitting. Some transceivers have a 'time-out timer' function which limits the time of a transmission which provides such a fail-safe. If that is not available some other form of timer control should be implemented on the remote amateur station particularly with solutions using remote access to a desktop computer controlling the radio equipment. An additional issue with some desktop computer remote access solutions is that a service provider holds the access credentials. Clearly if such a method to access a desktop computer which is then used to control a radio transmitter the amateur is not ultimately in control of their amateur station!

Remote Station Technologies

There is a vast range of remote amateur station technological implementation and indeed a wide range of associated costs. In this section I will address the basic requirements and some of the available technological components that will permit many radio amateurs to successfully set-up and operate a remote amateur station within the UK licence conditions. I have limited the scope of this chapter to solutions that use TCP/IP networking, both private networking and the Internet.

In the simplest case all that is required is a voice channel from (RX) and to (TX) the remote station and some way of exercising control over the transmitter or transceiver functions. Early remote station technology involved audio carried over telephone lines and customised DTMF signalling to control the station. Today this is best implemented by some form of VoIP (Voice over IP) for the audio and tunnelling of the transceiver's computer or control interface commands again over IP. Most of the solutions described in the rest of this chapter will follow this model.

Remote Receivers

Some readers might just want to implement a remote receiver particularly to take advantage of a lower noise location. However, there is little technical difference apart from control and security constraints between a remote RX and a fully remote amateur station!

If you have an old PC or can get a small processor to work with Simon Brown G4ELI's SDR Radio. The URL for this is in the reference section at the end of this chapter. This is by far the simplest and easiest remote receiver scheme to implement. SDR Radio has the advantage of a choice of audio compression techniques to reduce the upstream bandwidth required which might be an issue with low data rate Internet connections and some radio links on the amateur bands. Almost any sort of SDR (Software Defined Radio) hardware including the simple RTL Dongle with an HF converter works well. I have used a Funcube Dongle with good results.

One technology that might come to mind is the WEB SDR where users can access radio receivers over the Internet with a simple browser. Indeed, there are a number of WEB SDR receivers located in the UK. Although I have run a WEB SDR server for my own access I would not recommend it for a beginner as it can be rather hungry in its consumption of network bandwidth as the number of users goes up. I used it with an Ettus SDR with a modified WEB SDR script to scan 50-150MHz in several non-continuous slices so that I could have a layered waterfall display on a browser to spot when sporadic-e propagation was climbing towards 144MHz.

If you do not want to run a PC on your remote site, consider the rebooting issues, there are fairly simple hardware solutions for a remote receiver using a PC at the control or user end. If you have a receiver or transceiver with a RS232 interface it is fairly easy. To do this you could use a Wiznet WIZ110SR card to tunnel the RS232 commands over Ethernet (Internet). These Wiznet cards are available from RS and many other suppliers for around £36 and often less. As you will read later on, this technology works well with rotators and other station components.

At the control end use of a virtual serial port is recommended, which is accessed from standard radio control software for instance, Ham Radio Deluxe (Ham Radio Deluxe - full paid-for software subscription that expects a serial connection to a local radio. I use Virtual Serial Port (VSP) software, from Eterlogic. There are many alternatives available here where the user has a wide choice of free and paid-for VSP software packages. For a remote receiver you only require audio in one direction, the author recommends use of VoIP hardware at the radio end, paired with a standard media player at the control end. There are a number of audio over IP cards available that use VoIP protocols. An audio over IP solution tested by a number of amateurs for this application is manufactured and sold by the Italian company Futura Elettronica costing around □86. This company's web pages show how the card can be used with VLC player running over the control PC. Indeed, if you did not want to use the control PC for the audio you could use another suitable configured card with an audio amplifier.

Full Remote Station Operation

Many manufacturers of radio equipment have implemented their own turnkey solutions for remote operation. However, all of these solutions are for a limited range of their own, usually top-end, transceivers.

Icom's RS-BA1 software(Icom RS-BA1 remote control software allows a user to remotely control a number of high-end Icom transceivers directly and others using a PC base server at the radio end. Although on the face of it this is a simple remote-control solution the configuration seems to be somewhat complex. The radio, server and user all require registration and from the literature it is unclear how user security is maintained over the three UDP ports used for communications. The minimum upstream bandwidth is somewhat higher than with the RemoteRig solution described later which might be an issue with simple ADSL or limited bandwidth radio connections.

Yaesu have introduced a hardware unit the SCU-LAN10(Yaesu SCU-LAN10 remote control unit which enables the FTDX101D and FTDX101MP to be remotely used over an IP LAN or the Internet when used together with the SCU-LAN10 software.

If a PC is available at the radio end then a remote desktop solution (with or without audio) might seem like an easy means to achieve remote station working. As commented earlier I would question how this satisfies the terms of the UK amateur radio licence, particularly in terms of who controls the state of the amateur station especially when the communications link fails.

RemoteRig

Many amateurs will not want to configure individual cards and two sets of PC software and look for a simple all in one solution. Such an all-in-one solution has been available for over ten years from RemoteRig the RC-1258 which comes as two black boxes, one at the radio end and one at the control end.

The RemoteRig RC-1258 is a multi-functional modem designed specifically for remote amateur station operation. It contains bi-directional VoIP for RX and TX audio controlled by the PTT function. It provides a keyer for CW operation although I have used a RC-1258 with a straight key! It contains a control channel which can be

configured for a very wide range of common amateur transceivers and one or two, dependant on the transceiver in use, additional RS232 channels that can be used for rotators or other station ancillaries. Although the preferred network connection is wired Ethernet the RC-1258 units have an extra cost option of Wi-Fi.

The main user interface for both the control and radio unit is a computer browser which communicated with an embedded web in each unit. Both units need to be similarly configured so that they can communicate with each other. Finding a web server on a local LAN can sometimes be difficult so RemoteRig have available to download a simple setup manager tool which enables users to configure the network including Wi-Fi connection. Once the web server on the RC-1258 units is visible on a PC browser the rest of the configuration is relatively easy. The RC-1258 user manual, downloadable from the RemoteRig site is extensive and covers detailed configuration for nearly all transceivers supported.

After both units are configured it is wise to test the system on a local LAN. Connecting a PC client of remote radio head to the control RC-1258 unit and your chosen radio transceiver to the radio RC-1258 unit. You will need to the radio unit an IP address on the LAN, to ensure that all is working prior to configuring the network firewall at the radio site and attempting to connect over the Internet. All remote radio solutions will require some

Fig 10.1: RemoteRig RC-1258 pair of black boxes.

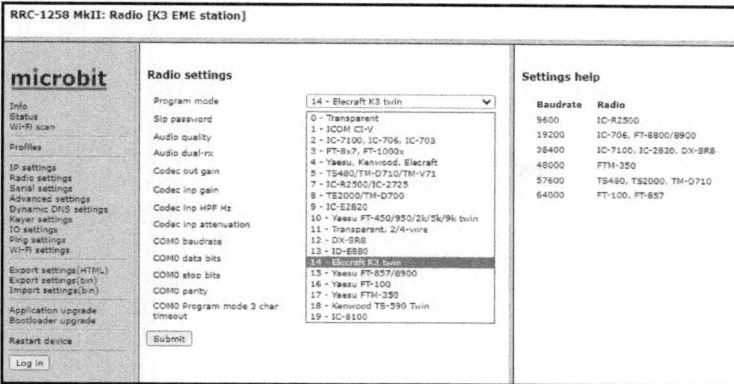

Fig 10.2: An example of the RemoteRig RRC-1258 user configuration, pull down menu.

network configuration and of the firewall at the radio site. These will be described in a later section of this chapter dealing with networking.

Radio and Control System

For many amateurs the choice of transceiver has already been made and they will be looking at how to provide remote access to their existing station. In this case it comes down to finding out what compatible remote solutions exist which will mostly be using a software radio client. If you choose to use the RemoteRig RC-1258 boxes there is a wide choice of software computer clients to control your radio. I have used Ham Radio Deluxe, both subscription and the free versions because I like the HRD log and that it has simple rotator software control. Commander which will control a vast range of transceivers has been recommended by others. Kenwood ARCP software offers proprietary solutions for quite a number of different transceivers including the older TS2000 and FT480. FTBCAT software is available for a number of Yaesu transceivers. There are also many other transceiver PC control programmes which can be used in conjunction with the

RemoteRig RC-1258 system of indeed other RS232 over IP solutions.

If you are only an occasional remote user and are happy using software control your transceiver RemoteRig have a simpler remote client solution the RRC-Micro PC-Client which is a plug-in USB device plus associated software. I have used the RRC-Micro when on short business trips when transporting more equipment would have been difficult. Indeed, although still in a beta version there is an RRC-Nano App for Android devices which allows your Android mobile phone or tablet to remotely operate your station.

Although I have experimented with various USB based knobs and tracker balls with software control programmes to facilitate tuning of the transceiver, I could not really train myself to this method of listening around the amateur bands. Fortunately, the second transceiver that I tested was a Kenwood TS2000 and after testing it with a pair of RemoteRig's RC-1258 and the RC-2000 remote head unit I never looked back. Although there was noticeable lag when remote tuning in less than

Fig 10.3: Kenwood RC-2000 remote working with RemoteRig RC-1258, Huawei MiFi 4G access and a PC interface for digital modes.

100Hz steps was attempted it still gave me a working solution for 50, 144 and 432MHz DX and casual HF working. Similarly, others have been highly satisfied with remote operation of the Kenwood TS480 using remote operation with its detachable control console. A number of Yaesu transceivers with detachable front panels, FT857, FT100 can be operated in a similar manner using the RemoteRig RC-1258.

I use a PC interface for digital modes which is a fairly standard design using transformers to isolate the PC audio and a FTDI USB to RS232 interface to provide the PTT signal to the RemoteRig RC-1258. Back in 2012 I upgraded my remote station to provide remote azimuth and elevation control of the rotators

Fig 10.4: Remote operation using an Elecraft K3-0 (early model in standard case).

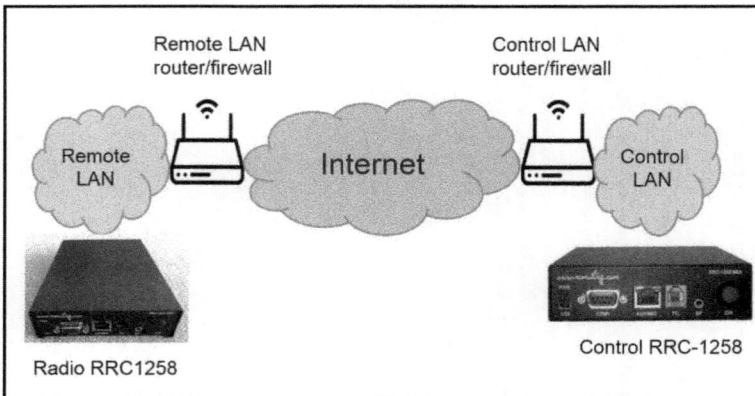

Fig 10.5: A pictorial representation of LAN and Internet domains typical of remote operation networking.

to enable me to work 144MHz, digital, moonbounce from almost anywhere. I have used this combination both as a home remote station, of course without the Huawei MiFi, and for demonstrations at radio clubs, presentations and on holidays for nearly ten years.

I had thought that the use of the Kenwood RC-2000 remote head unit in conjunction with the TS-2000 would be hard to beat. I was wrong. A number of transceivers allow 'locked' operation with another identical transceiver using the RS232 control interface. I have tested this with a Yaesu FT1000MP and indeed the Kenwood TS2000 but found the results when remotely controlled via the RemoteRig RC-1258 prone to delays and occasional errors. I then tested an Elecraft K3 transceiver using another K3 as the controller. It seemed to work perfectly with very low tuning lag. Around this time the engineers at Elecraft must have been testing the same thing as soon after my tests they released the K3-0. A box containing all of the control circuitry but no RF components. The ergonomic improvement with the K3 running with the K3-0 was staggering. The tuning was far more responsive with lower lag than all of the Yaesu, Kenwood and Icom transceivers I had tested. It seems that Elecraft chose to implement their computer interface in a different manner to the others which has resulted in superb remote operation capability.

Indeed after 5 years of remote operation using an Elecraft K3 and VHF/UHF transverters I was satisfied that I have achieved an almost perfect remote home station solution. For digital modes I still prefer to use audio from the Kenwood TS2000 which has a flatter response as it only uses DSP filters in the receiver. It is reassuring to read that the new Elecraft K4 is capable of full remote operation although whether there will be an equivalent of the K3-0 is unknown.

Internet Operation

All remote amateur station projects involve an amount of sorting out the network configurations of the various systems involved. For many radio amateurs this will be beyond their previous technical experience. in this chapter I will outline the basic issues but will not go into step-by-step guidance which is included in most equipment user manuals. Most IT Professionals, which includes many radio amateurs will have enough networking knowledge and for those who are in doubt they should ask a friend to help.

Most remote station solutions involve the use of: a LAN (local area network) at the radio end (remote LAN), the public Internet and a LAN at the control end (control LAN). The boundary or border between each domain is usually a router which directs network traffic between domains and often provides a policing or firewalling function on the traffic flowing between the domains.

The RemoteRig RRC-1258 user manual is highly detailed with step by step guides for most radio transceivers and network configurations. Other remote solutions will require similar configuration of the radio side interface and the router/firewall on that LAN. On the LAN side where the radio is located it is preferable to allocate this unit a fixed IP address on the LAN. This simplifies the configuration of the router/firewall where traffic from the LAN accesses the Internet. On the firewall a number of Network Address Translation (NAT) rules will need to be configured in order to allow inbound Internet communications to be routed through to the radio RRC-1258. This is generally 3 UDP ports and if you so desire a TCP port for the web interface. Here I would recommend changing this web port number to something more obscure than 80 to reduce the number of automated hacking attempts from the Internet. There is then the issue of how traffic from the Internet finds it way to your router on the radio site. The easiest way is to get a fixed IP address assigned to your Internet facing network connection. This can be fairly cheap and simple, my ISP charged me a one-off fee of £5 but others can be very expensive, charging a monthly fee and many do not offer this service. If you are unable to use a fixed IP address for your radio site some form of dynamic Domain Name Service (DNS) is required to direct traffic to your radio site. RemoteRig run a dynamic DNS service to assist users of the RC-1258. The user manual has instructions in configuring the RC-1258 units to use this service.

One way of getting around some of the issues of dynamic IP addresses and Internet routing using virtual private network (VPN) was proposed in a Radio Communications article; The Cambridge Community Remote Station Project (The Cambridge Community Remote Station Project, Radio Communications January 2016 P20) This VPN method gets over a lot of issues concerning IP addresses and public Wi-Fi access for the control system, but comes with a penalty of increased technical complexity.

The use of a mobile data service for the control end can often be more reliable and for occasional remote operation such as holidays etc may provide more reliable and less contested Internet access compared with hotel and other public Wi-Fi services. My own preference is for a Huawei MiFi such as is shown in **Fig 10.3** with its own data SIM card although one can use a personal hotspot configured on a mobile phone. I also have a patch type external antenna for the Huawei MiFi which gives better data speeds in poor coverage areas. When setting up the network firewall at the radio end I have often used a mobile data connection, to enable the RC-1258 control system to have its own separate Internet access on the radio site to test that the firewall configuration is correct.

The use of a 3G, 4G or 5G mobile data service for the remote station, apart from it being very expensive, can often be very difficult to get going. This is because most mobile network operators use one or several layers of network address translation. This does not present a problem for the control end of a remote-control solution but can make mobile

Internet access exceptionally difficult at the radio end. In practice this means that the IP address allocated to the radio connection can often change faster than dynamic DNS services can catch up. The solution is to use a specialised router which makes an outbound network connection, usually to set-up a VPN, to connect to the control PC.

Non-Internet Operation

Rather than using the Internet it is also possible to use some form of private networking, often over radio to provide distant connectivity. Operation using the control equipment hosted on the remote LAN provides a good way of checking that the network equipment is properly configured. This can also allow convenient operation from any room in the house! Here the use of encrypted Wi-Fi is perfectly acceptable as it falls within the licence exemption for Wi-Fi technologies. For slightly longer distances across private property Ethernet cabling or optical fibre is usually the cheapest networking solution.

Many radio amateurs have set-up long-range data links to enable them to operate remote amateur stations. In many cases it is not practical or within a reasonable budget to get an Internet connection to the remote station, so a long-range data link is the only viable solution. The equipment used is often designated 'Long Range Wi-Fi' but rarely falls within the UK licence exemption for Wi-Fi. Such data links usually use more power and high gain external aerials and must be operated under the terms of an amateur licence although the frequencies in use, at 2.4GHz and 5.7GHz can be the same as licence exempt Wi-Fi. When you set up such a data link under the conditions of the amateur licence encryption of the radio traffic is not permitted. The most popular equipment used for amateur data links of this sort is the Ubiquiti Bullet M series of transceivers which are designed for outside operation. Most amateur data links use mesh dishes which can provide up to 24dBi of gain. Where even a mesh dish does not provide enough gain for the link budget there are Wi-Fi

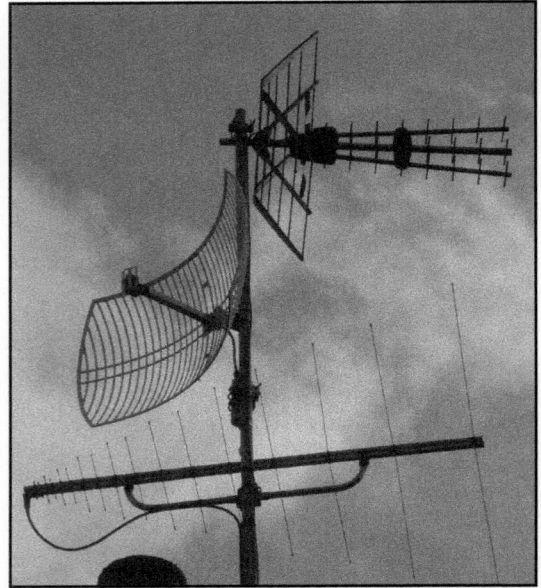

Fig 10.6: Long range, 2.3GHz 18Mbps, data link with dish and modified Ubiquiti data transceiver mounted under a TV antenna.

amplifiers available, with up to several watts output for difficult or long paths.

Without modification these long-range Wi-Fi transceivers can operate outside the permitted amateur bands so care must be taken configuring the frequencies used and operational bandwidth. Even then operation on spectrum used by Wi-Fi can lead to interference on the receive side and sometimes with normal Wi-Fi equipment. A far better solution is to move the operational frequency into spectrum that is not shared with Wi-Fi. We are fortunate that an amateur band lies below the frequency range used by 2.4GHz Wi-Fi. Some of the extended channels that can be accessed by firmware modification of the Ubiquity M2HP transceivers are in the part of the amateur band below 2.350GHz. The detailed technology of microwave data links falls beyond the scope of this chapter. If you have high, low latency and jitter bandwidth available, such as a 20+ Mbps private data link then you can control and backhaul I + q packets from a network SDR. I use an Afedri AFE822X that I operate in such a manner. Some details are in a presentation by the author at the 2016 RSGB Convention: "Microwave Links for Remote Stations" and many details can be found on the Internet.

If you can restrict the traffic to the remote station to <150+Kbps there are several other options. Including New Packet Radio on 146MHz or 437MHz. There are also quite expensive OFDM data link modules that can be set up in the 430MHz band and operated within the terms of the amateur licence.

Other Remote Equipment

If you own an older multi-band HF linear amplifier it is unlikely that you can run it remotely without a major project to implement your own remote-control system. Some newer solid-state linear amplifiers have their output filtering controlled by the transceiver and do not require separate remote control apart from on-off control. With these and single band untuned VHF/UHF linear amplifiers, output and VSWR monitoring is highly desirable. In the next section I will describe a web enabled power meter.

With a number of more modern auto-tuning HF amplifiers there is again a 'black box' solution from RemoteRig available, the RC-1216H. This is a small embedded web server that controls the linear amplifier. Current coverage is for a range of Expert amplifiers, Acom 2000 and Elecraft KPA500. The same unit can also be used to control the configuration of SteppIR tuneable antennas.

I have used a RC-1216H in conjunction with an Acom 2000 amplifier for over six years flawlessly on most of the HF bands. Here, I will add that control of the Acom 2000SW remote antenna switch is also included allowing multi-band operation without a separate remote antenna switching arrangement. The display on a web browser is very similar to the display on the Acom 2000 control panel making remote operation close to identical to being in the radio shack.

One issue not mentioned so far is the one of building in an ability to switch on and off the mains supply. Adding a web-controlled

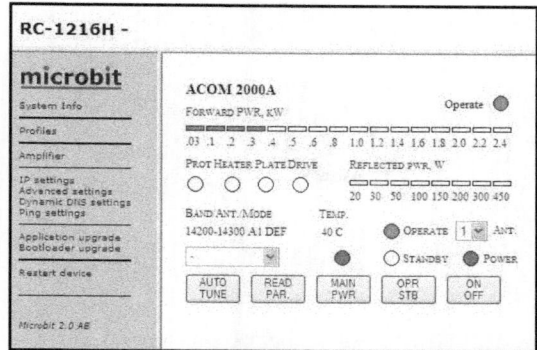

Fig 10.7: Screen shot of the web browser display of Acom 2000 controlled by RC-1216H.

switch with a high current mains contactor is an ideal way of ensuring that the remote station is made safe when not in use. There are many suppliers of Web controlled switches or relays many offering additional functions.

The RemoteRig Webswitch 1216H has five relay control outputs which can be addressed remotely with a web browser. The Webswitch 1216H also has the flexibility to control rotators using a web interface either using a rotator with a RS232 connection or using a specific 1216L rotator control card. Some users prefer a web interface to the rotator rather than configuring serial ports and using a software rotator package such as PstRotator.

I have used Web controlled relay cards from Kmtronic KMtronic web-controlled relays are a cheaper solution if you want to integrate a printed circuit board with a web-controlled switch into a larger system control unit.

There are many other rotator alternatives to using the solutions from RemoteRig most costing far less if you are prepared to take a little more time in integrating a number of separate components. I have used Wiznet WIZ110SR RS232 over Ethernet cards with rotators with a built-in RS232 interface. When the rotator does not have a RS232 interface this can be easily added with a kit from EasyRotor.

On the control PC, virtual serial ports can be used in conjunction with an RS232 over programming task to duplicate most of the functionality provided by the RemoteRig boxed solutions. The Remote QTH system (Remote QTH server uses Linux code over a Raspberry PI has many possible configurations and could easily form the basis of a remote station project.

I have not attempted to duplicate technology that can be purchased off the shelf but one component that was not readily available was a web-based power meter to interface with a dual-head Bird Thruline wattmeter. This project uses web sockets from an embedded web server running on a small Freescale 'Freedom' card and a JavaScript download to configure the display. The web sockets design was chosen to have the minimum update data rate to allow continuous operation even with restricted upstream bandwidth from the remote station. The greatest problem with this particular project was the analogue input amplifiers to match the Bird Thruline elements!

Lastly but always recommended is the use of web browser-controlled IP webcams. Before I had a web-based power meter to monitor the output of my linear amplifier I was forced into using such a webcam as a monitor. It is also useful to have such a webcam looking out at the antennas to ensure that all is in order. With careful selection of the definition

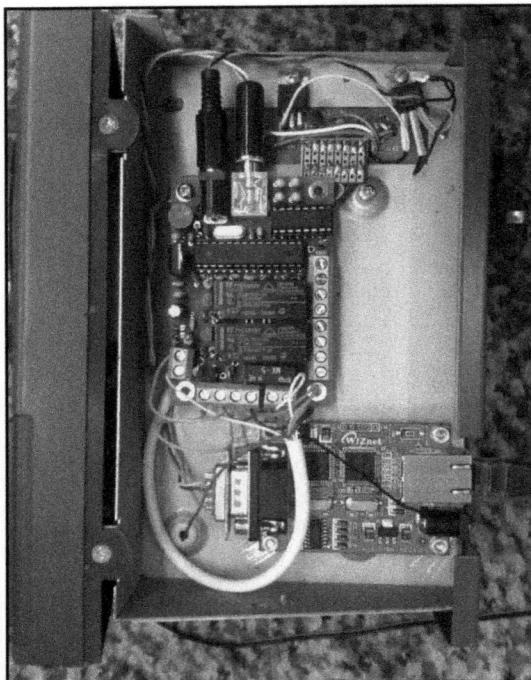

Fig 10.8: A Wiznet and Easy Rotor card fitted in an older Yaesu G1000 rotator controller.

Ethernet solution to drive rotator control software. I have usually used Eterlogic VSP software(Eterlogic VSPE Software driving PstRotator software running on my remote-control PC. Whilst I realise that PstRotator can interface directly over TCP/IP I have found issues when the network connection to the remote station has been congested whereas Eterlogic VSP always seems to seemly recover. I have now had ten years failure free operation with sometimes five rotator systems all using Wiznet cards at the remote station, Eterlogic VSP and PstRotator software on the control PC.

Many of the remote station hardware components can be homebrewed using small embedded processors. It being a moderate software

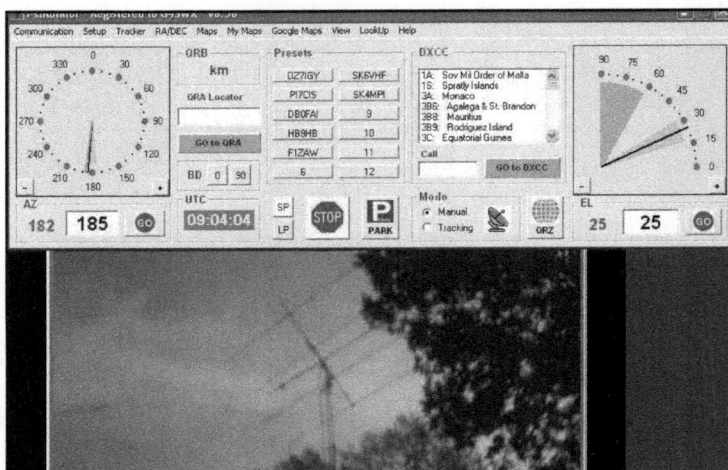

Fig 10.9: Screenshot of PstRotator display for both azimuth and elevation of a 144MHz remote EME system.

and frame update rates reasonable pictures can be viewed on a web browser at the expense of 100-200Kbps of traffic per camera.

Operational Considerations

It is important that all radio amateurs consider insurance of what is often an expensive investment in kit. As mentioned earlier, ensuring the security of an amateur station is one of the terms of the UK amateur licence. However, few amateurs fully consider the wider aspects of this issue. For the physical security aspects an easy way to look on this is how you might meet the criteria of an insurance company. This is particularly important if the remote station is located in an outbuilding rather than a domestic residence. Typical considerations are a five-lever lock on all doors, window locks and possibly an alarm or CCTV monitoring and recording system.

Fire safety is often overlooked when amateur operation is carried out in the presence of all of the station equipment. Small burn-outs or even minor fires can be dealt with quickly. For home stations I would always recommend having a small fire extinguisher in the shack. With remote operation the consequences of a small fire can be far worse as it is likely to remain unnoticed until it has spread further.

Remote amateur stations have not been looked upon positively by some contest organisers and awards bodies. Some seem to think that operating a remote station somehow gives the operator an unfair advantage. Remote operation within the terms of the UK amateur licence is simply having the control and the audio from the station extended to another location. This does not change the fact that the station has been established in the UK under the terms of the Ofcom issued amateur licence which are common to all UK radio amateurs. It would be wise for anybody running a UK remote amateur station to check on relevant contest rules and award conditions.

Conclusion and Looking Towards the Future

Fully featured remote access to amateur stations has been around for over 10 years and the technology is still advancing. Improvements in Internet connection speeds have reduced the issues with latency and the up-stream bandwidth required. Some of the more recent amateur transceivers with Ethernet interfaces have made remote operation without using a host PC possible. However, the increasing replacement of RS232 serial interfaces with USB connectivity can make simple remote control more difficult to implement.

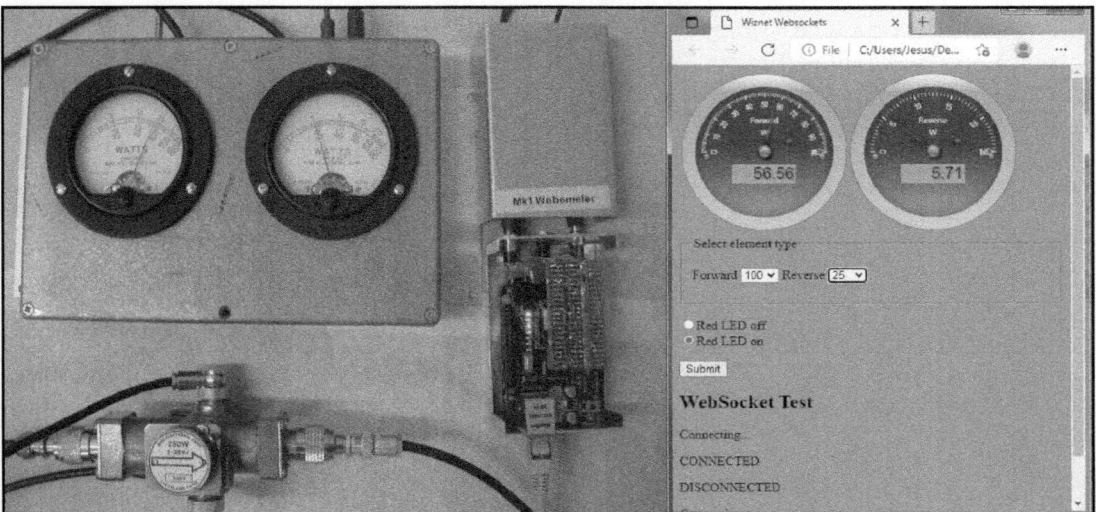

Fig 10.10: Homebrew 'Webometer' remote dual-head Bird wattmeter.

Some form of streaming of data between the control and the remote station will always be needed. To achieve the minimum data-flow some sort of finite state-machine set up over the remote connection looks like the best method. Such a state machine could easily be built using small computing platforms such as the Raspberry-PI and Arduino and could cope with legacy transceiver control interfaces and newer protocols as they arrive.

The most ubiquitous user interface is a web browser but that would have ergonomic limitations as has PC client software discussed earlier. For digital modes a web interface to WSJT-X or similar software packages, coupled with user authentication, could allow operation from any smart phone or PC almost anywhere. Already almost all station accessories can be remotely configured using a web browser. I have mentioned some and designing and building most of the others should be relatively easy for those skilled in small computing platforms.

In this chapter I have attempted to give a broad-brush technology overview to cover most of what is currently available in 2021. One of the remote station implementations that I have not seen so far is a fully featured remote amateur club station which allows more than one club member to use the station and others to listen in remotely to whoever is operating. Hopefully such technology will be developed in the next few years.

References

https://www.ofcom.org.uk/data/assets/pdf_file/0027/62991/amateur-terms.pdf

Simon Brown G4ELI's SDR Radio https://www.sdr-radio.com/console

WEB SDR receivers located in the UK: http://www.websdr.org/

Wiznet WIZ110SR RS232 over Ethernet card: https://uk.rs-online.com/web/c/computing-peripherals/networking-connectivity/interface-adapters/

(Ham Radio Deluxe - full paid-for software subscription: https://www.hamradiodeluxe.com/ Virtual Serial Port (VSP) software, from Eterlogic (Eterlogic virtual serial port software: http://www.eterlogic.com/Products.VSPE.html).

Futura Elettronica (Audio over IP hardware card: https://www.futurashop.it/filo-diffusione-audio-su-rete-ethernet-7100-ft1185m)

VLC player (VLC Player: https://www.videolan.org/)

(Icom RS-BA1 remote control software: https://www.icomjapan.com/lineup/options/RS-BA1_Version2/)

Yaesu SCU-LAN10 remote control unit: https://www.yaesu.com/downloadFile.cfm?FileID=16697&FileCatID=42&FileName=Network%20Remote%20Control%20System%20SCU-LAN10_Software%20Update%20Information.pdf&FileContentType=application%2Fpdf)

RemoteRig RRC-1258: https://www.remoterig.com/

Commander transceiver control software: http://www.dxlabsuite.com/commander/

Kenwood ARCP programs: https://www.hamradioandvision.com/kenwood-arcp-remote-programs

FTBCAT Software for Yaesu transceivers: https://ftbcat.software.informer.com/2.0/

Ubiquiti Bullet M: https://www.ui.com/airmax/bulletm/

New Packet Radio Data Link: http://www.m0ahn.co.uk/nprabout.html

RemoteRig RC-1216H: https://www.remoterig.com/wp/?page_id=1010

PstRotator from YO3DMU: https://www.qsl.net/yo3dmu/index_Page346.htm

EasyRotor kits: https://www.schmidt-alba.de/eshop/index.php?cPath=1

KMtronic web-controlled relays: https://www.kmtronic.com/

Eterlogic VSPE Software: http://www.eterlogic.com/Products.VSPE.htm

Remote QTH server: https://remoteqth.com/remoteqth-server.php

11
D-Star

by Dave Thomas, MW0RUH & Andrew Barron, ZL3DW

The first part of this chapter has not been updated as the original author was not available and a new author could not be obtained. However, there has been the addition of information about DRM and System Fusion which was kindly provided by Andrew Barron from his book *"Work the world with DMR Digital Mobile Radio Explained."* , Febuary 2021

Digital Smart Technologies for Amateur

Radio (D-Star) is a communications standard, not a brand name, which is not limited to one manufacturer. It was released in 2001, having been developed and funded by the Japanese Ministry of Post and Telecommunications.

The standard is published by JARL, but it is an open system, which means that any equipment complying with the standard can use it. It is a two part communications system, the first being formed by radio-to-radio transmissions - either direct or via a stand alone repeater - and the second deemed to be the 'spine' (or backbone) of the system is by integrating with the Internet via 'gateways' to the wider amateur community. The D-Star standard also controls the way in which the signal is relayed, by converting voice to and from digital format. This digital exchange takes place by the use of the AMBE (Advanced Multiple Band Encoding) codec ('codec' being short for coding/decoding).

D-Star Modes

D-Star carries digitized voice and digital data in two different ways, there being a combined voice and data mode (DV) and a high speed data only stream (DD).

Although data and voice are carried at different rates and are managed in different ways, they are transmitted as packets.

The AMBE codec can digitise voice at several different rates. D-Star uses 2.4kbps (bits per second). In addition, AMBE adds information to the voice data that allows the codec at the receiving end to correct errors in the transmitted stream. The result of the overhead is that the digitized voice stream carries data at a rate of 3.6kbps.

Simultaneously to the digitised voice, DV Mode (low speed data mode) can also carry 8-bit digital data at 1200bps. This data is unmodified when transmitted, so it is up to the operator's software to manage the flow of data whilst it is being exchanged.

When operating in DD Mode (high speed data mode), the voice signal is unused and all packets are dedicated to the use of digital data. Transmitted data is sent as raw data at a rate of 128kbps. Like DV mode, this is transmitted with no modification, the flow control being undertaken by the software package chosen by the user. In DD mode the net data rate is comparable to or better than a high-speed dial-up Internet connection.

How D-Star works

Being a packet based protocol, D-Star data is processed and packaged using the required data and additional information. Packets are sent in their entirety and are processed as a group by the receiving station.

D-Star is a one-way protocol, so no response is required from the receiver to ac-

Header				Data							
Sync	Control flags	ID data	Check-sum	Voice frame	Digital	Voice frame	Digital	. . . .	Voice frame	Digital	Check-sum

Fig 11.1(a): Structure of DV mode.

Header				Data
Sync	Control flags	ID data	Check-sum	Ethernet packet

Fig 11.1(b): structure of DD mode.

knowledge that a packet has been received. The reason D-Star does not require acknowledgements is because, as previously stated, it has error detection and correction built into the datastream.

The structure of the DV and DD modes are illustrated in Fig 11.1. Each consists of a header and data segment.

D-Star utilises a common method of using one protocol to send data formatted according to another protocol. In the DV packet, voice data is contained in short segments (frames) which are formatted according to the AMBE protocol. In the DD packet, the data is formatted using the Ethernet protocol. This process of putting data from one protocol 'inside' another is called encapsulation.

The illustration of the packet structure is broken down in Fig 11.2 as follows:

Sync Frame
Bit Sync is a standard pattern for GMSK 1010 modulation used by D-Star. Frame Sync is '111011001010000' - a unique bit pattern in D-Star packets.

Control Flags
Control flags are used to direct the processing of the packet.

Flag 1 Indicates whether the data is control data or user data, whether communication is simplex, repeater, set priority, etc.

Flag 2 Reserved for future use as identification data.

Flag 3 Used to identify the version of D-Star protocol being used, sothat as new functions are added the receiver can apply them

Identification Data
Received Repeater Callsign - Callsign of the repeater that is to receive the packet

Send Repeater Callsign - Callsign of the repeater sending the packet

Counterpart Callsign - Callsign of station that is to receive the data

Own Callsign 1 - Callsign of the station that created the data

Own Callsign 2 - Callsign suffix information

P-FCS Checksum
A checksum is used to detect errors. The P-FCS checksum ID is computed from the flag and ID data.

How D-Star Corrects Errors in Digital Voices
D-Star uses two methods of combating transmission errors:

1. Error Detection codes are used to detect errors. These codes only tell the receiver that the data is damaged

Sync		Control			Identification					Error
Bit sync	Frame sync	Flag 1	Flag 2	Flag 3	Received Repeater Callsign	Sent Repeater Callsign	Counterpart Callsign	Own Callsign 1	Own Callsign 2	P-FCS

Fig 11.2: D-Star packet structure.

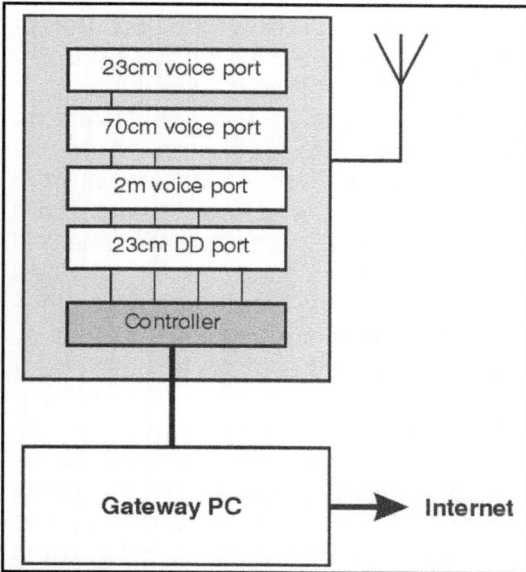

Fig 11.3: Icom D-Star Repeater.

D- Star System Layout

A D-Star repeater can be built according to the keeper's requirements and be active on several bands with the same callsign unlike an analogue repeater. Fig 11.3 illustrates a full repeater stack setup. As you can see from the illustration, a fully loaded D-Star repeater can be constructed with four ports. It can also be built with any combination of the four ports.

Assuming a complete setup, a user would be free to access the system on any of the bands avail-able within the repeater. Across the world it has been decided that wherever possible the 'A' port will carry 23cm (1.2GHz) voice traffic, the 'B' port 70cm (430-440MHz) voice and slow data, the 'C' port 2m (144-146MHz) voice and slow data, and the D port 23cm (1.2GHz) high speed data .

For the system to be able to take advantage of the capabilities of D-Star, the repeater requires a broadband connection to the Internet to 'Gateway' the completed project. It should also be mentioned that a gateway connection requires specific Linux software to allow full operation on an

or corrupted. D-Star checksums follow the CRCCCITT Standard.

2. Error correcting codes contain information about the data. Because the codes are sent with the data (to enable correction at the receive end), they are called Forward Error Correcting or FEC codes. FEC codes contain enough information for the receiver to repair most damage.

Both the DV and DD data packets in Fig 11.1 use the P-FCS checksum in the header, but the DD packet also contains the Ethernet data packet checksum at the very end.

With the DV packet data segment, each AMBE digitised voice frame contains its own FEC code to allow the receiver to repair errors.

DV Digital data frames are not protected, relying on the applications to detect and correct errors.

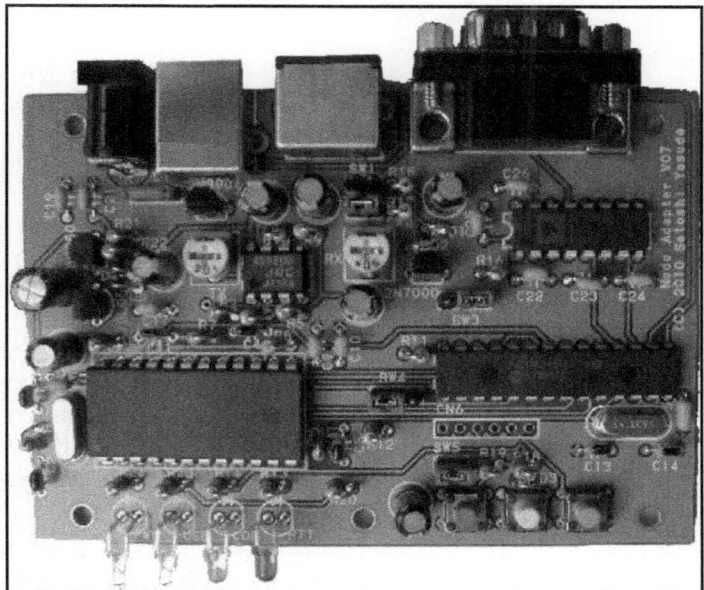

Fig 11.4: Satoshi node adaptor board.

Fig 11.5: The DV Dongle.

Fig 11.6: DV Access Point (DVAP).

Icom factory built system. A server-based PC would give the system the best results, although a good spec PC will work well.

Several homebrew options are available in simplex (licenced in the UK as MB6xx callsigns) and in full repeater mode (GB7xx callsigns). These use either Satoshi Yasuda's DV Node Adaptors (see Fig 11.4).

Any one of these boards can be connect-ed to an analogue radio with a 9600 packet data port to create a passage for the packet to be passed via the adaptor to the system. You will, however, still require a D-star radio to access these nodes, as they do not provide a platform for an analogue radio to enter the system (which some believe they do). There are also several software applications around to operate these nodes, either via Windows or Linux platforms. Some amateurs have coupled these hotspot adaptors to dummy loads, to provide an access point solely for their use at home. This enables access with a D-Star handheld on very low power.

For those with property or antenna restric-tions, or if travelling regularly, a DV Dongle - Fig 11.5 -is the best choice for operation. This is a device with the AMBEcodec installed and works in a similar way to Voice over Internet Protocol (VoIP).

Any PC running the DV tool software along with the DV Dongle and a suitable headset can access D-Star from anywhere using WiFi or fixed broadband connections.

Fig 11.6 shows the DV Access Point (DVAP). It transmits 10mW of RF on the 2m band (frequency of your choice) and is con-nected (similarly to the DV Dongle) to a PC and dedicated software, which then permits the use of a D-Star radio to access the net-work anywhere within range of the DVAP, via the PC's Internet connection.

Operating D-Star

Most of the problems for new users seem to be associated with correctly configuring their radio to operate the mode. This because, unlike analogue where you switch on your radio and tune to a specific frequency and begin to operate, D-Star is menu driven and its main requirement is to ensure that certain information is included in the header (Fig 11.7) to guarantee that your conversation is

Sync		Control			Identification					Error
Bit sync	Frame sync	Flag 1	Flag 2	Flag 3	Received Repeater Callsign	Sent Repeater Callsign	Counterpart Callsign	Own Callsign 1	Own Callsign 2	P-FCS

Fig 11.7: D-Star header.

UR	Sets who you send to
RPT1	Sets the local repeater call and band
RPT2	Sets call routing - Local or Distant
MY	Your callsign

Fig 11.8: These four fields are sometimes known as the 'Ohms Law' of D-Star.

heard and routed to where you want it to go.

D-Star has a more complicated configuration. The operator is required to input information into their radio to correspond with the identification section of the header.

This part of the radio configuration, shown in Fig 11.8, is sometimes described as the 'Ohm's Law' of D-Star. It contains four fields of information corresponding to the identification contained within the header. All D-Star radios require this information set in an exact manner to make it work (simplex excluded, where only CQCQCQ [UR] and your callsign [MY] is required).

We are going to set some information into the menus, with an explanation of some likely information that must be contained in it to make that all-important QSO.

Note: As with an analogue repeater, your radio would need to be programmed with the frequency of GB7xx. The output frequency would have to be set and the 'shift' set to the designated split, so that it transmits and receives on the appropriate frequencies. There is no CTCSS used in D-star. In the setup shown in Fig 11.9;

UR=CQCQCQ allows the general call to be placed. It is also the most used field in the identification process and can be varied according to where and who you want to speak to.

RPT1 setting selects GB7xx and port B (could be A,C or D, if available) as the part of the repeater you wish to communicate through.

RPT2=NOT USE tells the controller not to route the call to the Internet, but to remain in the local vicinity.

MY: Is your own callsign and identifies you as the user.

UR=CQCQCQ
Tells the gateway not to route your call to any particular repeater or station. A general CQ to all stations.
RPT1=GB7xx-B
Indicates that you wish to communicate using the repeater GB7xx and on port B (generally 70cm)
RPT2=NOT USE
Prevents your call from being routed to the internet.
MY=
Your callsign

Fig 11.9: Typical information set in a radio menu fields.

A note about the setup in Fig 11.9. When RPT2= is set to 'NOT USE' or is blank, if the repeater is 'Gateway' connected to the network (linked) any user on the wider system will not hear the station who has RPT2= set this way. For those who appear to listen to one sided conversations on D-Star, it is the RPT2= setting that is responsible for this.

There is one more rule that needs to be applied which has not been mentioned so far and that is D-Star commands.

For the network to recognise that a stream wishes to pass, it must receive a command to execute what is required by the user.

We have seen that GB7xx is required and we have selected port B, but for the system to accept this command of using port B the 'B' itself must be in the 8th character position when programming RPT1.

This 8th position is going to come up throughout the rest of this chapter, as it is the basis for all Linking and Unlinking (see Fig 11.10). So to 'Gateway' to the Internet, the only change required is to change RPT2 from 'NOT USE' to the callsign of the repeater in use, adding a 'G' (Gateway) to the callsign. As shown in Fig 11.11. With this, plus the UR, RPT1, RPT2 and MY callsign data in place, you should successfully be in position to undertake that all-important first QSO on D-Star.

Computers in Amateur Radio

Character position	1st	2nd	3rd	4th	5th	6th	7th	8th
Callsign setup	G	B	7	X	X	Blank space	Blank space	B

Fig 11.10: Menu set with the important 8th character (RPT1).

Character position	1st	2nd	3rd	4th	5th	6th	7th	8th
Callsign setup	G	B	7	X	X	Blank space	Blank space	G

Fig 11.11: Menu set with the important 8th character (RPT2).

Wider System Operation

The wider operation of the D-Star network relies on use of the Internet and the network connections within it. Any D-Star user must register their callsign on the system to link or unlink to the available connections within D-Star. Registration is normally done with the user's local D-Star repeater. This enables the local repeater keeper/group to verify the identity of a user

Reflectors

A reflector is a 'hub' or conference room where users congregate. The most up-to-date list can be found on the Internet at www.dstarinfo. com

Reflectors normally have three possible connections (modules). There is no strict rule as to which of the three modules (A, B or C) is used for any specific use, it is the decision of the reflector owner.

Fig 11.12 shows the Dashboard of Reflector 5, the London-based main UK reflector and the connections made on some of the modules.

DV Dongles and some simplex gateways/ hotspot users and non-Icom repeaters will also be shown within the DV Dongle section on the dashboard page. Having established that the wider network involves the use of reflectors, it

is now useful to explain further how to do this connecting and disconnecting of repeaters to reflectors - something that an analogue system does using a DTMF code to connect or disconnect to IRLP or Echolink nodes. D-Star uses the identification portion of the header to do this and it is done by changing the UR: from 'CQCQCQ' to 'REF005AL'. The command 'L' (Link) is again in the 8th position, as shown in Fig 11.13. All other menus remain as previously set, RPT1: with your local repeater and port setting, your RPT2: with your local repeater and 'G' Gateway setting and the MY: your own callsign set.

Once connected, the UR: menu needs to be switched back to 'CQCQCQ', to undertake your QSO whilst connected to your now chosen reflector.

Once your conversation is completed you may wish to disconnect your local repeater. To do this you use the unlink command, simply a 'U' for unlink. See Fig 11.14.

This will bring your local repeater into Standalone mode for local use (remember again to switch the UR: menu back to 'CQCQCQ').

Some repeater keepers auto link their gateways in a semi-permanent way. This is to encourage users to use the network by keeping traffic on their gateway.

Those repeaters with semi permanent con-

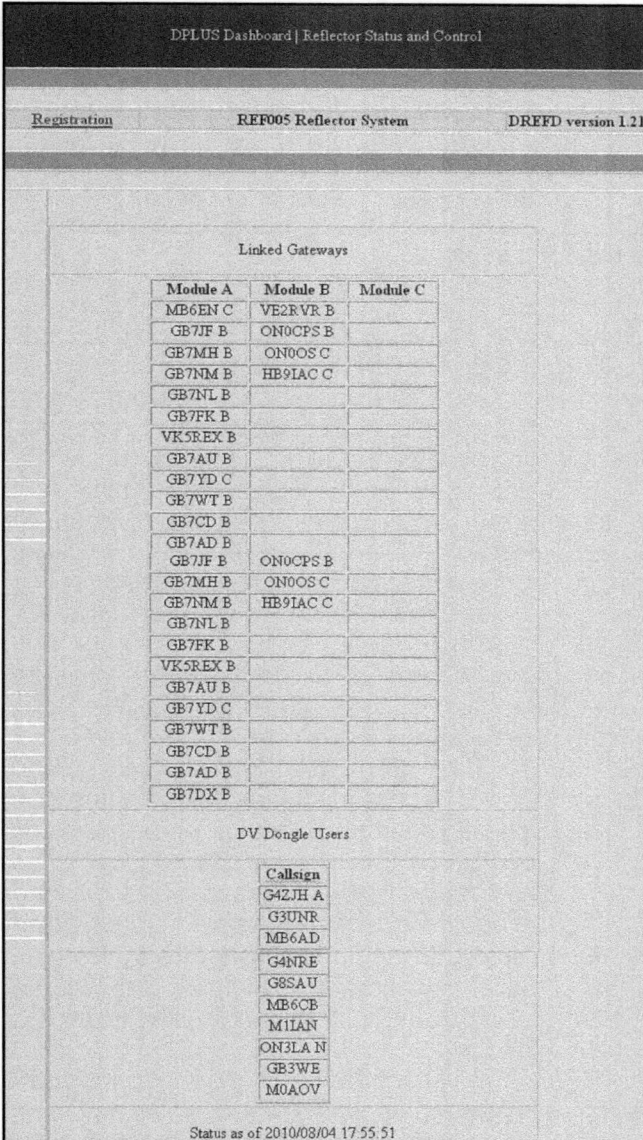

DPLUS Dashboard | Reflector Status and Control

Registration   REF005 Reflector System   DREFD version 1.21

Linked Gateways

Module A	Module B	Module C
MB6EN C	VE2RVR B	
GB7JF B	ON0CPS B	
GB7MH B	ON0OS C	
GB7NM B	HB9IAC C	
GB7NL B		
GB7FK B		
VK5REX B		
GB7AU B		
GB7YD C		
GB7WT B		
GB7CD B		
GB7AD B		
GB7JF B	ON0CPS B	
GB7MH B	ON0OS C	
GB7NM B	HB9IAC C	
GB7NL B		
GB7FK B		
VK5REX B		
GB7AU B		
GB7YD C		
GB7WT B		
GB7CD B		
GB7AD B		
GB7DX B		

DV Dongle Users

Callsign
G4ZJH A
G3UNR
MB6AD
G4NRE
G8SAU
MB6CB
M1IAN
ON3LA N
GB3WE
M0AOV

Status as of 2010/08/04 17:55:51

Fig 11.12: Reflector 5 Dashboard.

nections have default scripts installed which automatically disconnect after a period of inactivity. This brings them 'home' to their semi-permanent location.

It is good practice to drop the link to the system when you wish to speak locally with a station on the same repeater, because it frees the reflector for other users.

There are other ways of using the D-Star network (repeater to repeater, direct callsign routing). The UR: menu is key to these operations. Once the RPT1: and RPT2: fields are set to your local gateway and stored in a memory in your radio (along with your own callsign), it is just the UR: field that will determine the success of your routing and use of the network. Once you become accustomed to the methods described here of connecting to the network, the rest should come easily at a later stage.

Other Uses of D-Star

D-Star can simultaneously send data whilst streaming voice, so you can talk to other users whilst typing messages to them via the keyboard of your PC. The only additional requirement to carry out this activity is the use of an interface lead to/from the radio.

Radio amateurs across the globe have become involved with the provision of software applications to exploit this data capability and

Character position	1st	2nd	3rd	4th	5th	6th	7th	**8th**
Callsign setup	R	E	F	0	0	5	A	L

Fig 11.13: Change UR menu to connect to Reflector 5A.

Character position	1st	2nd	3rd	4th	5th	6th	7th	8th
Callsign setup	Blank space	Blank space	Blank space	Blank space	Blank space	Blank space	Blank space	U

Fig 11.14: Change UR menu to unlink from the system.

it is widely used by radio emergency teams (especially in the USA) for weather nets and so forth.

Some software packages allow low speed file transfer, pictureexchange and e-mail capability. Examples of these are D-Rats, D-Chat, D-Star TV and D-Star Comms,

D-Star can transfer your GPS position to servers such as APRS.fi, it can help you locate another user on GPS, tell you which direction you need to travel and how far away that station is, then effectively home you into them by following the direction arrow on your radio's front panel to their beaconing signal. Imagine this in an emergency situation, where medical care maybe urgently required! Even their latitude and longitude is displayed on your radio. The 23cm DD mode in particular has endless possibilities.

Computer Connection

D-Star radios are quite complex and it is well worth investing the extra money to purchase the necessary interface lead and software dedicated to the transceiver of your choice.

Programming can be quite tedious and a long process to learn, but many local D-Star groups have ready-made files that can be dropped into the radio via the software, saving many hours of hard work. Looking through these files on your PC will also give you a better understanding of how the mode really works.

Developments in Digital Radio

Since the previous edition was written, there has been a lot of development on this side of the amateur radio hobby. D Star is now no longer the only standard. In this section we will explore developments within D Star itself and the newer digital technologies such as Fusion and DMR.

This US Trust system is a production network which discourages any experimentation. D-Plus was added to the G2 system by Robin Cutshaw, AA4RC, as a transport mechanism for an operator's callsign information to be distributed across all of the connections to the system and allowed for direct routing to users by replacing the normal 'CQCQCQ' in the UR: menu to the callsign of the person you wished to make contact with. The problem with this was that if the station was mobile and travelling across several nodes, the information was slow in being spread across the network by D-Plus. This often led to failed attempts to hold a successful QSO, because by the time the routing information had been propagated the user had switched to another connection so the call was being routed to the wrong node. Improvements have now been made to improve the speed in which the information is relayed across D-Plus.

The secondary network began to experiment and improve the way in which D-star was being utilised. This was being led by the amateurs in Germany, but the biggest changes to the way things were being influenced came about from Jonathan Naylor, G4KLX, who started in 2009 with repeater software development (soundcard based for D-Star and FM) purely because there was no home-brewing going on. His work began to revolutionise how the network could be accessed. He approached the US Trust and Robin Cutshaw to assist in developing their network, but was rejected.

The German amateurs made use of Internet Relay Chat (IRC), which is used by millions of people and thousands of organisations to communicate, share, play and work with each other on IRC networks around the world. IRC was utilised to manage the database that pushes operators' callsigns in a similar way to which D-Plus was, the difference being that it was almost instantaneous so it cut down on the time it took to move the data across the nodes around the world. This transport mechanism was named as ircDDB (Internet Relay Chat Database).

Jonathan worked by the request of the German team on this method of transportation and developed a program called ircDDB Gateway, which was open source software. The first release of this was in September 2010 and it very quickly became adopted for a number of repeaters. GB3IN was the first to use this repeater software and also the first one introduced to ircDDB. This allowed access to this new network and it started a rapid trend for homebrew and experimental equipment to be use

IrcDDB has always encouraged openness, whilst most other groups do not. It is this openness that has made this network the success it is today, because - after all - this is what our hobby was intended to be about, self-learning and experimentation.

Not only was Jonathan developing the ircDDB Gateway software, but also Digital Voice and PC repeater open source software. It also include other protocols like DMR, P25, etc. Some of this software is compatible for use on analogue as well as digital repeaters and networks. The name has recently changed to OpenDV (Open Digital Voice) and the project is available at http://opendv.berlios.de

In relation to D-Star, due to this open source approach, some variations of the original software began appearing in other parts of the world. In Canada Ramesh Dhami, VA3UV, formed the FREE STAR* system, which appears to utilise this software. There was another developer also active in the early days that developed software; Scott, KI4LKF.

So the X-reflector system began to flourish, with all the individual experimenters doing their thing. 'Dextra' reflectors were added to this secondary network, although for what purpose I cannot say. It appeared to just spread the network and give more choice but made it slightly more difficult to get people together as the users spread across the network. The same could be said with the US Trust system, which was also growing to some extent.

In the meantime more hardware was becoming available, with PCBs and manufactured boards from the US, Holland and Poland entering the fray; but with Jonathan's software rapidly progressing the need for a board was becoming redundant as his software required little more than a soundcard and a PTT circuit to add to a compatible analogue radio to access the network as a simple node, either as a personal hotspot or as a licensed simplex gateway. As I write this, this soundcard and a rather 'over the top for the purpose' Vellman board is my hardware set up on MB6BA.

Meanwhile the German amateurs decided to progress with a 'DV-RPTR' board and their own gateway software, which is getting rave reviews as a solid and reliable piece of hardware for D-Star. This DV-RPTR board offers future add-ons for more functionality and is still developing.

In early 2012 the protocol coding was rewritten to combat the dropout issues mentioned earlier in the chapter. .The re-coding changed the way in which the header information and packet data was read. This changed D-Star for the better and as a result a DCS reflector was introduced and branded the second generation of D-Star repeaters. Any person who did drop out due to fringe coverage on a node miraculously came back mid over, which was never possible prior to DCS.

DCS reflectors became the 'in' thing and users migrated in their droves to join in this new breed of usable system. In the UK many

	Last Update (UTC):	Registered:	Activated:	Online:	Onl/Reg:	%total:
ircDDB:	2012/08/23 12:33	951	946	598	62.9%	69.1%
US-Trust:	2012/08/23 12:02	847	842	531	62.7%	61.4%
Common:	2012/08/23 12:02	412	410	264	64.1%	30.5%
ircDDB only:	2012/08/23 12:33	539	536	334	62.0%	38.6%
no ircDDB:	2012/08/23 12:02	435	432	267	61.4%	30.9%
Total:	2012/08/23 12:02	1386	1378	865	62.4%	

Fig 11.15: IrcDDB Live.

repeater keepers added IRCddb to their repeaters, although some still resist.

What's the Difference to Those Who Utilise IRCddb Nodes Over Non IRCddb Systems?

The users on the G2 network have just that - G2 connected Nodes. Users of the ircDDB network have access to the second generation

The introduction of DCS reflectors also did what D-Plus said couldn't be done, which is the ability to utilise DTMF (Dual Tone Multi Frequency) tones. The introduction of this has simplified the way in which D-Star can be accessed and makes it easier for the user. Please remember though that (at the time of writing) if the Gateway you are accessing doesn't have ircDDB, DTMF is not possible.

So How Does DTMF Make This Simpler?

Using the system as described earlier in the chapter, D-Star makes use of the 'UR: menu' in your radio. IrcDDB configured gateways have this information stored at the gateway, so by using the pre-determined DTMF codes for your wanted connection it puts you where you want to be. Consequently your radio only needs the repeater callsign and module letter in RPT1 and RPT2 set to your local gateway, plus your callsign in the MY: menu, and the UR: menu set to CQCQCQ. By simply pressing the PTT on your radio and utilising a DTMF microphone you key in the module number and you will be automatically disconnected from the repeater's current connection and reconnected to your chosen link. This includes connecting to G2 based systems too. Non-ircDDB repeaters will still need individual memories to access each connection on the G2 network and you will not be able to access DCS / X-Reflector or Dextra from a G2-only system.

ircDDB Live

home docu status live sources support download register site notice

13:01:37	********* ____	CQCQCQ__	MB6AD__C	MB6AD__G	·		13.7s S:5% E:0.0%	00 00 0
13:01:38	DL8FCE__ IC92	CQCQCQ__	DB0HAM_B	DB0HAM_G	·	berhard DOK C 28	0.2s S:100% E:12.5%	00 01 0
13:01:43	LB6YD___ IC92	CQCQCQ__	DB0HAM_B	DB0HAM_G	·	berhard DOK C 28	0.2s S:100% E:12.5%	00 01 0
13:01:45	VE3XZT__ ____	CQCQCQ__	VA3ODG_B	VA3ODG_G	·	LV DALE MOBILE*6C	TX on	00 00 0
13:01:50	********* ____	STN055__	IR7UBA_C	IR7UBA_G	·	·	0.3s S:0% E:0.0%	00 00 0
13:01:51	SM6ZDO__ ____	CQCQCQ__	DB0HAM_B	DB0HAM_G	·	CQ DStar QSO Party	0.2s S:100% E:12.5%	00 01 0
13:01:51	MW0SGR__ ____	CQCQCQ__	GB7CD__B	GB7CD__G	REF003_C	JAMES +OBILE WALES	4.7s S:0% E:0.0%	00 00 0
13:02:24	DG8BCN__ I56_	CQCQCQ__	DB0CHV_B	DB0CHV_G	XRF026_B	Tanja, mobil, 2820,	1.0s S:11% E:0.0%	00 00 0
13:01:53	VA3BL___ 2820	CQCQCQ__	VA3PMO_B	VA3PMO_G	XRF005_B	·	·	00 00 0
13:01:55	DL8FCE__ RALF	CQCQCQ__	DB0HAM_B	DB0HAM_G	·	Dongle Gateway V 1.0	0.2s S:100% E:12.5%	00 01 0
13:01:57	MW0SGR__ ____	CQCQCQ__	GB7CD__B	GB7CD__G	REF003_C	JAMES MOBILE WALES	TX on	00 00 0
13:02:00	********* ____	CQCQCQ__	IR1UBR_B	IR1UBR_G	DCS002_A	·	0.3s S:13% E:0.0%	00 00 0
13:02:00	SM4XEX__ E880	CQCQCQ__	DB0HAM_B	DB0HAM_G	·	HANS FALUN SWEDEN	0.2s S:100% E:12.5%	00 01 0
13:02:07	SA7ADV__ 2820	CQCQCQ__	SK7RQX_C	SK7RQX_G	DCS010_B	Kristian Perstorp	3.1s S:0% E:0.0%	00 00 0
13:02:03	G8SAU___ ____	XRF044AL	GB7NU__B	GB7NU__G	·	LINK-XRF044A	0.2s S:0% E:0.0%	00 00 0
13:02:04	DL8FCE__ RALF	CQCQCQ__	DB0HAM_B	DB0HAM_G	·		0.2s S:100% E:12.5%	00 01 0
13:02:08	DO5BFB__ ____	CQCQCQ__	DB0ELD_B	DB0ELD_G	DCS010_V	WWW.I57.DE FLOrIAN	7.6s S:0% E:2.9%	00 00 0
13:02:09	DL8FCE__ E880	CQCQCQ__	DB0HAM_B	DB0HAM_G	·	HANS FALUN SWEDEN	0.2s S:100% E:12.5%	00 01 0
13:02:15	SM4XEX__ E880	CQCQCQ__	DB0HAM_B	DB0HAM_G	·	ANS FALUN SWEDEN	0.2s S:100% E:12.5%	00 01 0
13:02:16	********* ____	CQCQCQ__	MB6AD__C	MB6AD__G	·		TX on	00 00 0
13:02:49	DG8BCN__ I56_	CQCQCQ__	DB0CHV_B	DB0CHV_G	XRF026_B	Tanja, mobil, 2820,	3.1s S:0% E:0.0%	00 00 0
13:02:19	DL8FCE__ RALF	CQCQCQ__	DB0HAM_B	DB0HAM_G	·		0.2s S:100% E:12.5%	00 01 0
13:02:20	IW1DGK__ IC91	SM6ZDO__	IR1CJ__C	IR1CJ__G	DB0HAM_B	CQ D-STAR QSO PARTY	1.1s S:0% E:0.0%	00 00 0
13:02:24	IV3ZJU_P IC92	CQCQCQ__	IR3UBK_C	IR3UBK_G	·	IV3ZJU Antonio [PN]	0.2s S:66% E:0.0%	00 00 0
You don't see your call sign here? Please read: http://ircddb.net/live-vis.html								
13:02:27	IZ5GTN_P PORT	CQCQCQ__	IR5AR__C	IR5AR__G	·	·	0.4s S:0% E:9.7%	00 00 0
13:02:32	IV3ZJU_P IC92	CQCQCQ__	IR3UBK_B	IR3UBK_G	·	IV3ZJU Antonio [Pn]	0.2s S:66% E:0.0%	00 00 0

Fig 11.16: DCS5 active connections dashboard.

The great thing about DCS is that there are up to 26 available connections per reflector. This gives reflector owners more scope to provide you, the user, with more choice.

A protocol seems to have been agreed for each DCS reflector to allocate the 'A' module for worldwide contacts, therefore all DCS reflector 'A' modules are linked together. Regardless of whether you go to DCS001'A', 002'A' or 013'A', you will be linked to the same connection. Another neat thing with DCS is that (in most cases) module owners have allocated dedicated chat channels; so if you make contact, say, on the UK national module of DCS005'B' and you would like a long ragchew with your mates, you can QSY to modules DCS005'O', 'P', 'Q' or 'R'. You can also visit the London area on DCS005'L' or the Midlands on the 'M' module, or maybe in the evenings Wales and the West on DCS005'W', or switch to other areas of the world such as USA (DCS006) or Australia (DCS014). Many Euro-pean countries also have their own reflectors, so if you are proficient at a foreign language you can communicate or practice your skills in these areas of the network.

With this new network comes new ways of following the data and Fig 11.15 shows the live feature page which can be accessed at www.ircddb.net Here systems that utilise ircDDB have data that corresponds with each PTT on the network and offers the users a location to see where activity is.

All active connections to each of the available DCS reflectors can be found on dedicated Internet pages, which identify where and how long each connection has been linked. It also shows the last user of that connection (see Fig 11.16).

A further Internet dashboard shows the different parts of that country that can be accessed on each of the modules, including it's DTMF and radio programming code for the UR menu. It also shows how many repeaters / connections are made on the module (Fig 11.17).

StarNet Digital Services

Further to the introduction of DCS and ircDDB, StarNet Digital (developed by John Hays, K7VE, and implemented by Jonathan Naylor, G4KLX) has been introduced to the network. Starnet Digital has the ability for

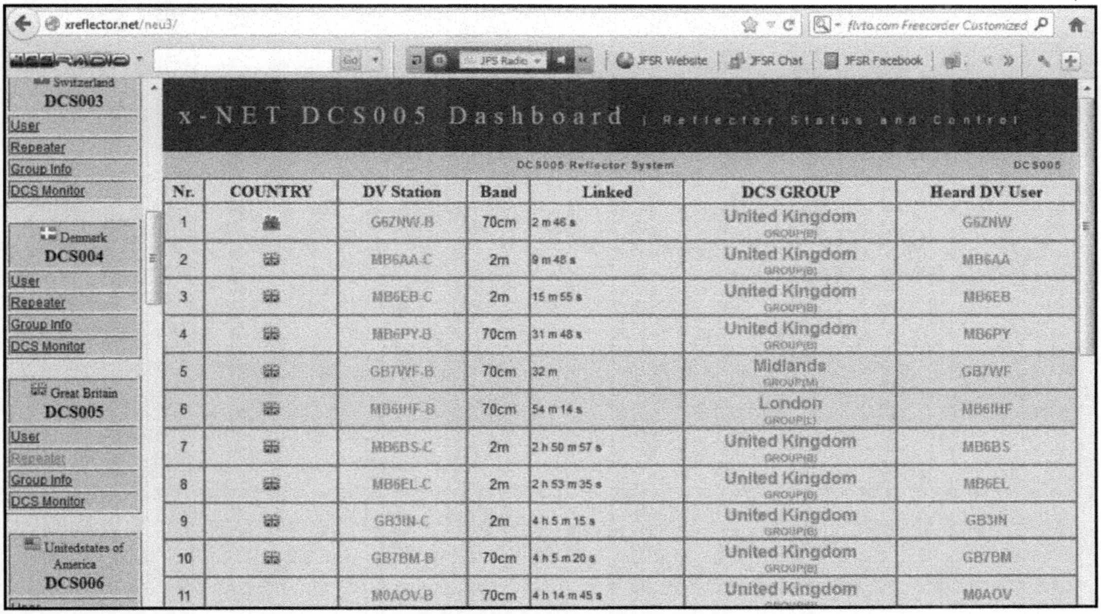

Nr.	COUNTRY	DV Station	Band	Linked	DCS GROUP	Heard DV User
1		G6ZNW-B	70cm	2 m 46 s	United Kingdom GROUP(B)	G6ZNW
2		MB6AA-C	2m	9 m 48 s	United Kingdom GROUP(B)	MB6AA
3		MB6EB-C	2m	15 m 55 s	United Kingdom GROUP(B)	MB6EB
4		MB6PY-B	70cm	31 m 48 s	United Kingdom GROUP(B)	MB6PY
5		GB7WF-B	70cm	32 m	Midlands GROUP(M)	GB7WF
6		MB6IHF-B	70cm	54 m 14 s	London GROUP(L)	MB6IHF
7		MB6BS-C	2m	2 h 50 m 57 s	United Kingdom GROUP(B)	MB6BS
8		MB6EL-C	2m	2 h 53 m 35 s	United Kingdom GROUP(B)	MB6EL
9		GB3IN-C	2m	4 h 5 m 15 s	United Kingdom GROUP(B)	GB3IN
10		GB7BM-B	70cm	4 h 5 m 20 s	United Kingdom GROUP(B)	GB7BM
11		M0AOV-B	70cm	4 h 14 m 45 s	United Kingdom GROUP(B)	M0AOV

Fig11.17: Area Allocations dashboard and codes for use.

user groups to be formed under a dedicated Starnet user Callsign, normally listed as 'STN' followed by the group number. This STN group can be joined by changing your UR:menu from CQCQCQ to the STN group callsign. A quick PTT logs you into the group and this then follows you across the network, like callsign routing. It just requires a quick PTT on each repeater you move to. StarNet Digital can be used by all amateurs to set up specific interest groups, such as HF contests, satellite users, SOTA groups etc.

The Raspberry Pi computer has successfully been introduced as the gateway repeater for some systems. The beauty of this is that it's cheap, consumes little power and is extremely small.

System Fusion and DMR

System Fusion (often known as Fusion) is a digital mode created by Yaesu in 2013. Like other digital modes, it is able to function well in environments where there is interference and other factors that make using a more traditional radio set up difficult if not, in some cases, impossible. It uses the C4FM FDMA (Continuous Four Level Frequency Modulation) standard.

One of the main differences between this and other dgital voice modes, is that it is backwards compatible with analogue FM.

There are four modes of operation.

1. V/D (VD is Voice Digital) mode (also known as Voice+Digital or Voice FR (VW) Mode). It uses the available bandwidth for high-fidelity voice operation to provide clear communication.
2. Highspeed data. This mode transfers data at the full rate of available speed.
3. Analogue FM mode. This mode is for backwards compatibility so existing equipment can be used with System Fusion.
4. Automatic mode select. This recognises if the signal being received is C4FM or standard FM. It then switches to the communication mode that match the received one. This makes operating

easier as the operator does not have to manually switch every time a different signal type comes in. Within this function there are some modes that the operator can choose.

These are:
1. Auto (ths is the automatic mode select),
2. TX manual. The RX/ TX mode is automatically selected from DN, VW, DW and FM to match the received signal but this can be changed manually by the operator.
3. TX FM fixed. In this mode, the RX mode is automatically selected from DN, VW, DW and FM but the TX mode stays as FM.
4. TX DN fixed. In this mode, the RX mode is automatically selected from DN, VW, DW and FMbut the TX mode stays as DN (Digital Voice Narrow).
5. TX VW fixed. In this mode, the RX mode is automatically selected from DN, VW, DW and FM but the TX mode stays as DN (Digital Voice Wide).

Yaesu are currently the only company that manufacture System Fusion radios. This information was found at http://systemfusion. yaesu.com/what-is-system-fusion/ and http://www.ws1sm.com/System-Fusion.html

DMR

What is DMR?

DMR stands for Digital Mobile Radio. It was developed as a way of reducing the bandwidth of the transmitted signal while improving the quality of the received voice transmission. DMR can transmit two voice channels on the same RF carrier, using the same 12.5 kHz bandwidth as a single FM repeater channel. Also, the digital voice signal is less prone to noise and flutter fading, resulting in excellent audio quality. This does come at the expense of needing a slightly higher received signal strength than an FM signal. Typically, the DMR signal from a repeater will sound "perfect" or it won't be received at all.

On Internet-linked services such as amateur radio DMR talk groups, packet loss or delay in the Internet traffic can sometimes break up or distort the audio, but this is not a failure of the DMR link.

In 2005 the Digital Mobile Radio Association was formed to promote compliance to the ETSI (European Telecommunications Standards Institute) standards among its member manufacturers. The association now includes more than 160 DMR equipment manufacturers. Motorola was the first company to market a range of DMR radios. These days, Hytera and Motorola are the largest producers of DMR equipment. Motorola uses the term MotoTRBO in place of DMR.

The biggest advantage of DMR systems is the 'open ETSI standard' which ensures that radios from all DMR radio manufacturers can be used on any DMR network. This is especially important for amateur radio DMR because it means you can buy any DMR Tier II radio, and it will work on your local DMR repeater or any DMR hotspot.

DMR supports private calling, group calls, short messages (SMS), GPS location, error correction, talk groups, better battery life than FM radios, and man-down safety features. Some DMR radios and networks also support APRS. The dual time slots make DMR difficult for scanners to decode, with the option to add encryption on Tier II and Tier III radios and network encryption on Tier III trunk systems. However, in most countries, encryption is not legal for amateur radio systems.

References: https://cwh050.mywikis.wiki/wiki/MotoTRBO and https://www.hytera.us/resources/dmr-tier-iii-3

The biggest disadvantage of DMR is that the system was developed for commercial radio networks where a user is given a radio that is pre-programmed with the channels and talk groups for the network. Although you can select or add talk groups, channels, and zones using the buttons on the radio, it is rather tedious. Realistically, to change the configuration, a new 'code plug' configuration file must be downloaded into the radio.

You need a channel for every talk group that you want to use, on every repeater or hotspot that you want to use it on. And each channel must be in a zone, or you cannot select it. You must also get the time slot and the colour code right. Programming your radio is very complicated. It takes a lot of time to get it right.

YAESU SYSTEM FUSION CONFUSION

System Fusion (YSF) has two network possibilities. You can connect to the Wires-X system which is operated by Yaesu. To do that you either need a YSF repeater that has a Wires-X interface, or a second radio connected to a Yaesu HRI-200 Wires-X modem which together acts as a hotspot.

This is an expensive use of a Yaesu radio, and you must use the Yaesu modem. The alternative is to use a standard YSF repeater or an MMDVM hotspot. This gives you access to some, but not all, of the Wires-X rooms, and some cross-linked talk groups on the DMR or D-Star networks.

Getting started

The first thing you need to do is check out what DMR repeaters are available in your area. It is a good idea to talk to DMR users at your local amateur radio club. If there are no DMR repeaters but you have good a good Internet connection, or you are willing to use your mobile data to provide a WiFi connection, you can buy a hotspot and use that to connect your radio to talk groups all over the world. Next, you should buy a DMR handheld or mobile radio. Take a look at some things to look out for, on page 5 in the previous chapter. I went for a mid-range Radioddity GD-AT10G. It has four power ranges up to 10 watts, a 3100 mAh battery, a large 200,000 contact list, 4000 channels, a programming cable, GPS and APRS. On the downside, it is UHF only which might not suit you if you want to use the radio on FM repeaters. It also includes radio keypad programming, which can be very handy if you go to a different region. The radio is the same as the AnyTone AT-D878UV except that the AnyTone radio is a dual bander. I also bought a TYT MD-UV380 because it is a popular model, and many other radios are programmed the same way. Once you have your radio, there is work to do before you can use it. At first, this seems rather daunting because of all the new terms you need to learn and having to download software and databases. But we will step through the process, and it will soon become second nature.

YOUR DMR ID

You cannot use any DMR repeaters or talk groups unless your radio has been programmed with your personal DMR ID number. This number is linked to your amateur radio callsign and will be used for all of your DMR radios and hotspots. There is only one registration source worldwide. It is at https://radioid.net/register or search for 'DMR ID registration.' Note that you must be a licenced amateur radio operator to get a DMR ID. Before you start the registration process make sure that you have a copy of your amateur radio licence or operating certificate (1 to 3 pages) in .gif, .jpg, .jpeg, .png, or .pdf format. I had an image from my LoTW application years ago, so that was fine.

TIP: If you buy a used DMR radio or hotspot you must ensure that you change its DMR ID to your DMR ID. Otherwise, you will be effectively illegally using someone else's callsign.If the party you are talking to has a large contact list loaded into their radio, it will display the previous owner's name, location, and callsign.

The process begins with the website asking for your callsign, name, and email address. It then sends you an email which you must open, but you do not need to click a link or reply. I guess it just gets a read receipt from the mail server. Then you can carry on and enter a password for your account. It must be at least 8 characters and contain a minimum of one symbol, one number, and one upper case letter.

Next, a new web page appears. It will usually be pre-loaded with your address. I guess they get that from QRZ.com.

You will have to enter any missing information. I had to put 'South Island' using the State/Prov dropdown list. Save that information to get rid of the red warning bar. The final thing you must do is upload a copy of your amateur radio licence or operating certificate (1 to 3 pages) in .gif, .jpg, .jpeg, .png, or .pdf format.

That makes the second red bar disappear and the application goes into a pending mode while your licence is manually checked. You should receive an email with your DMR ID number within a few days. My application was approved in four and a half hours. I was automatically issued with DMR IDs for my DMR radio and two hotspots.

Your DMR ID is used for all your DMR radios and your first hotspot. Subsequent hotspots get a two-digit 'ESSID' extension 01, 02 etc. to separate them in the network from the first hotspot. A DMR repeater can use your DMR ID with an extension, but most licenced repeaters have a different callsign issued by the licencing authority.

12
Automatic Packet/Position Reporting System (APRS)

Edited by Lorna Smart, 2E0POI

This is an edited version of the chapter written by Chris Dunn, G4KVI that was in the previous edition of this book.

Automatic Packet/Position Reporting System (APRS), developed by Bob Bruninga, WB4APR (www.aprs.org), is a lightweight system that allows users to transmit location and other data in single data packets. Normally, stations being tracked use GPS receivers to provide real time tracking data.

APRS uses existing packet Terminal Node Controllers (TNCs), which are small, low-cost micro-controller driven units containing modems to transmit standard AX.25 packets on a frequency of 144.800MHz at 1200 baud, but APRS can also be used over HF and satellite links.

APRS is intended as a short-range tactical system, but it can also be viewed over broad areas using Internet gateways. Such gateways can be run on low-cost computers and can relay the transmission of packets to and from the international Automatic Packet/Position Reporting System - Internet Server (APRS-IS).

As a multi-user data network, it is different from conventional packet radio in four main ways.

1. By the integration of maps and other data displays to organize and display data.
2. By using a one-to-many protocol, updating everyone in real time.
3. By using generic digipeating, so that prior knowledge of the network is not required.

4. Since 1997, a worldwide transparent Internet backbone, linking everyone worldwide.

Consequently, APRS turns packet radio into a real-time tactical communications and display system. Normal packet radio is useful in passing bulk message traffic (e-mail) from point-to-point, but it does not do well at real-time events where information has a very short lifespan and needs to get to everyone quickly.

Although APRS is mainly intended to be used locally, the Internet monitors APRS worldwide, but this is not the primary objective. However, like our other radios, how we use APRS in an emergency or special event is what drives the design of the APRS protocol. Although APRS is used the vast majority of the time over great distances and benign conditions, the protocol is designed to be optimized for short distance real-time crisis operations using RF.

APRS provides universal connectivity to all stations by avoiding the complexity and limitations of a connected network. It permits any number of stations to exchange data, just like voice users would on a voice net. Any station that has information to contribute simply sends it, and all stations receive it and log it.

How APRS Works
An APRS station broadcasts (beacons) a single packet of information to all stations in range. This packet usually contains GPS co-ordinates and other information. The

packet may be received and decoded by any station that can hear it and has suitable software or hardware. Digipeater (Digital Repeater) stations can also hear the packet and rebroadcast it based on rules in the digipeater software and commands that are integral to the packet. Packets that need to travel long distances can also be routed across the internet. What happens next is that the packet is re-transmitted by every

digi than can hear it, including any Internet Gateway (iGate). The packet is then re-transmitted by every digi that heard the first digi. The packet is heard by every APRS station in direct range of this second set of digis, including the first one.

The final step of the process, is that the packet is again rebroadcast by every digipeater in direct range of the second set of digis, including the original digipeater. Correct setting of the Unproto command is essential to control this and prevent channel overloading.

The fundamental principles of APRS, as described by Bob Bruninga, are:

- The system should provide reliable, real time, tactical digital communications.
- Use a 1200 baud network system operating as an Aloha random
- access channel (see below).
- You should hear everything nearby or within one digipeater within ten minutes.
- You should hear everything within your Aloha circle within 30 minutes.

An ALOHA channel provides access to a common communication channel from multiple independent packet transmitters by the simplest of all mechanisms. When each transmitter is ready to transmit its packet, it simply transmits the packet burst without any coordination with other transmitters using the shared channel. If each user of the ALOHA channel is required to have a low duty cycle, the probability of a packet from one user overlapping and thus interfering with a packet from another user is small as long as the total

number of users on the shared ALOHA channel is not too large. As the number of users on the shared ALOHA channel increases, the number of packet overlaps increase and the probability that a packet will be lost due to an overlap with another packet on the same channel also increases. Hardware such as a Terminal Node Controller (TNC), a GPS Receiver, a radio and a computer are used with APRS. Depending on what you are using the APRS for will depend on how much of this you will need.

If you want to see APRS stations, you'll need some software. There are software packages for most operating systems. Once you have decided on your hardware there are also some basic software settings you will need to know such as Secondary Station IDentification (SSID), latitude and longitude unproto address, beacon comment and status text

Practical Applications

To what practical application can APRS be put? The most obvious is realtime tracking. This can simply be an amateur using an APRS enabled transceiver or mobile phone to show their position on UIView maps. It is also widely used by RAYNET groups during events, to keep track of vehicles. An example would be to place APRS trackers in St John Ambulance or Red Cross vehicles. The ability to 'see' where vehicles are is a valuable tool when responding to an incident. It can also be helpful to 'see' in realtime where the 'lead' and 'tail' vehicles are, for example on a marathon event. The messaging facility is also very useful. The ability to send a message to a number of stations in realtime has endless possibilities.

Another exciting use for APRS is as a propagation indicator. Over the course of time you will come to know which stations you normally receive directly. As conditions change due to a 'lift', you will notice more stations appear on the map in the direction of the enhanced propagation. This feature can also be useful when installing a new aerial system.

Mention has been made of Internet servers. The information from these is collected and displayed on APRS.fi. This website shows a large amount of data from APRS stations, including tracks of mobile stations.

Mobile Phones

In recent years the use and availability of smartphones has exploded, the most popular being the iPhone and those running the Android software. There are several applications available which enable these devices to run APRS. These use the signal from the 3G / WiFi network and the inbuilt GPS to send location information to an APRS Internet server These applications enable the device to be used as a full feature APRS tracker. Although this may be seen by some as not amateur radio this is an exciting development enabling realtime tracking without the use of an expensive radio or computer. One again it shows how amateur radio can evolve and embrace new technologies.

13
Electromagnetic Compatibility

by Leo Ponton M0NNQ

Whilst the use of a computer in the shack has been shown as necessary for the implementation of many amateur radio facilities, such as logging or PSK operating modes, great care is required to avoid interference with the main purpose of radio communication. In a nutshell, the computer and all of its associated peripherals and interfaces must be electro-magnetically compatible. That is, the computer must not interfere with radio communication, and it must be immune to interference from the transmitters in the shack. This requires considerable care and consideration by the operator.

Equipment Issues

All items of electrical equipment are manufactured and tested to comply with international standards, which gives some degree of protection to the radio spectrum, and an assurance that devices/apparatus will be reasonably immune to interference from other apparatus.

The standard EN 55022/CISPR requires that any equipment shall not emit signals above limits that give general protection to radio services. In practical terms this means that emissions shall not mask broadcast radio services.

Many situations are catered for in the standards, but the following tables most closely typify the issues which are to be found in the average home and therefore radio shack.

Firstly, below 30MHz, all limits are specified by reference to 'conducted disturbances'. This is because of the difficulty

	Limits dB(uV)	
Frequency range	Quasi-peak	Average
0.15 to 0.50MHz	66 to 56	56 to 46
0.50 to 5MHz	56	46
5 to 30MHz	60	50

Note 1: The lower limit shall apply at the transition frequencies.
Note 2: The limit decreases linearly with the logarithm of the frequency in the range 0.15 to 0.5MHz.

Table 13.1: Limits for conducted disturbance at the mains ports

of making radiated measurements in this frequency range. It can be seen in **Table 13.1** that of class B ITE. quite a large signal in terms of dB(μV) is permissible. Note that 60dB above 1μV = 1 Volt. These signals do of course radiate!

Table 13.2 covers all connections to/from equipment, such as peripherals, interfaces

	Voltage limits dB(uV)		Current limits dB(uA)	
Frequency range	Quasi-peak	Average	Quasi-peak	Average
0.15 to 0.5MHz	84-74	74-64	40-30	30-20
0.5 to 30MHz	74	64	30	20

Note 1: The limits decrease linearly with the logarithm of the frequency in the range 0.15MHz to 0.5MHz.
Note 2: The current and voltage disturbance limits are derived for use with an impedance stabilization network (ISN) which presents a common mode (asymmetric mode) impedance of 150Ù to the telecommunication port under test (conversion factor is 20 log10 150 / I = 44dB).

Table 13.2: Limits of conducted common mode (asymmetric mode) disturbance at telecommunication ports in the frequency range 0.15MHz to 30MHz for class B equipment.

Frequency range	Quasi-peak limits dB(uV/m)
30 to 230MHz	30
230 to 1,000MHz	37

Note 1: The lower limit shall apply at the transition frequency.

Note 2: Additional provisions may be required for cases where interference occurs.

Table 13.3: Limits for radiated disturbance of class B ITE at a measuring distance of 10m.

and any type of data connection - all generally referred to as telecommunication ports.

This specifies the limits on 'common mode' signals on these ports. The use of the expression 'common mode' may not be generally understood, but it is an important concept in the context of interference. (See explanation box below). Once again it can be seen that the interference signal can be very large.

Finally we get to the situation above 30MHz, where radiated measurements are used, and the distance between the interfering source, and the 'victim' is set at 10m. In **Table 13.3** the signal strength is given in dB(μV)/metre.

Having established what is in the 'Standards', consider the practical situation that radio amateurs face when operating on the bands. Even the average HF communication receiver is able to discern signals down to a few microvolts, and at VHF/UHF signals down to 0.1μV are several dB above the noise and therefore quite readable.

Choosing the Right Hardware
Power Supplies/Motherboards/ Cables
Possibly the best advice on choosing a computer for the shack is to spend as much as you can possibly afford. There is no doubt that the EMC effectiveness is clearly reflected by the quality of the product. There is no need for the latest and fastest, with super graphics performance. Moderate performance will be satisfactory, unless of course you intend to use the machine for other purposes - but do think about what you will do with the machine and purchase accordingly.

Tower vs. Laptop Computers
Tower PSUs
Probably the biggest contributor to RFI is the power supply. These days all computer power supplies are Switched Mode units. These are notorious for radiating unwanted emissions. Even when they use 'best practice' they can be detected somewhere in the HF spectrum. The biggest single offence that manufacturers commit is to omit the mains input filter components. **Fig 13.1** shows an example of a power supply where the filter has been omitted at the manufacturing stage. Without opening the case it is difficult to determine whether these components have been fitted. However, it is possible with an optic-scope to take a peep inside through the rear fan opening. In any event, the absence of the filter parts will be

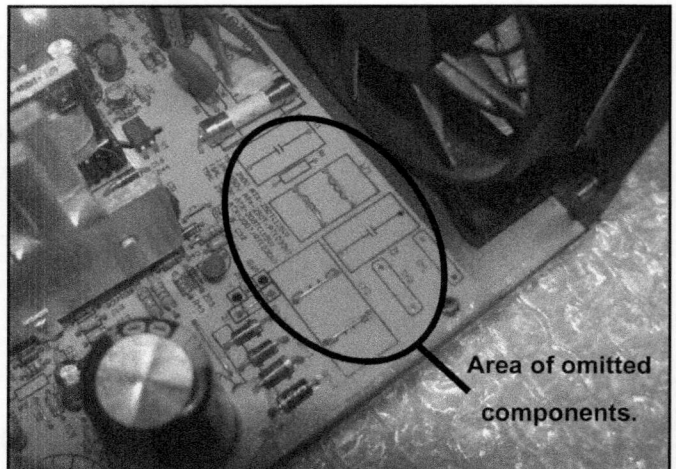

Area of omitted components.

Fig 13.1:Purchased from a High Street vendor - showing how the power line filter components were omitted in manufacture to save cost. This unit failed to meet the limits by more than 20dB. Area of omitted components.

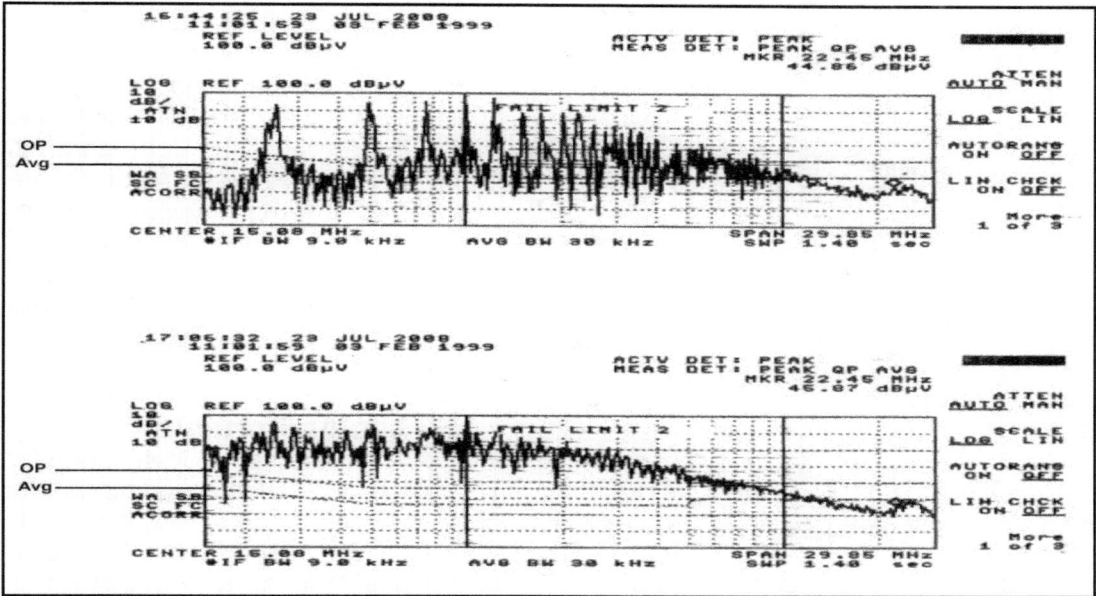

Fig 13.2: Plots of the emissions from an unscreened power supply (a) in standby, (b) operating with a 60 watt load on +12V.

obvious on an HF receiver when the computer is turned on. See the recommendations that follow! **Fig 13.2** gives typical plots of a second power supply (see **Fig13.3**), showing the emissions when in 'stand-by' mode, and when loaded with a nominal operating current. In this case the load was a passive resistance, avoiding any possibility of other emissions. The limit lines indicate the Class B standard limits.

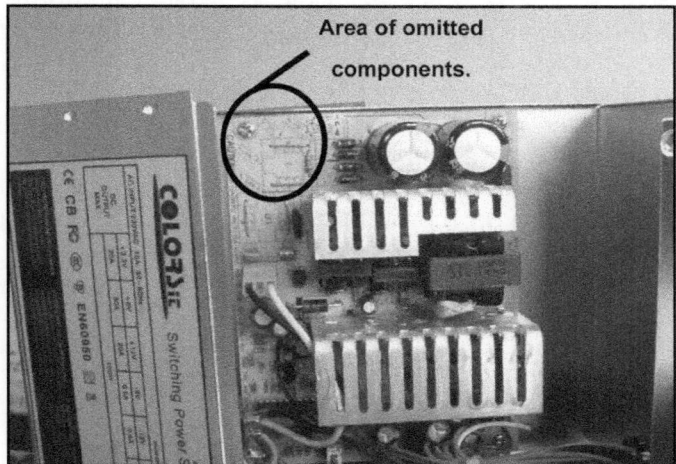

Fig 13.3: Another power supply in which the power line filter components were omitted in manufacture.

Cases and Cabinets
Build it Yourself vs. a Branded Model

There are as many cases available on the market as there are power supplies and the quality issues are exactly the same. These cases are often produced in large quantities by a variety of manufacturers, not necessarily the same manufacturer from one batch to another. They will have little or no knowledge of EMC design considerations and they have probably copied an existing case design and missed or economised on the important aspects of ensuring that all gaps close and all parts make good electrical contact when screwed together.

Many cases use fingering along closing edges to make contact with an opposing part, but frequently the fingering is found to be flattened, in which case it does not contribute to electrical connection.

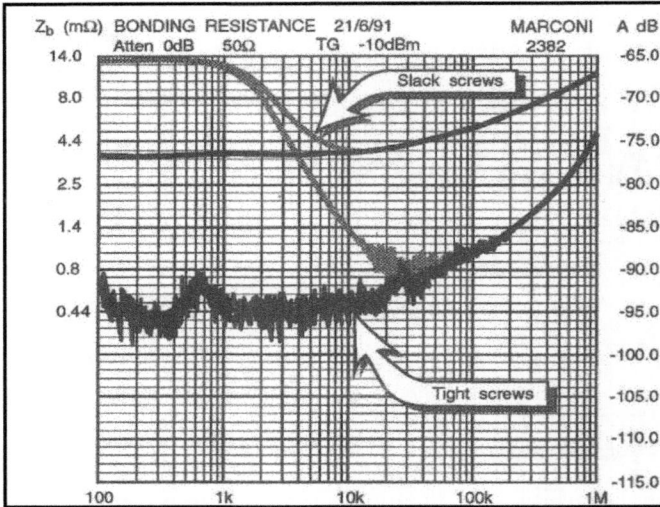

Fig 13.4: The bonding resistance of supply line filters.

Purchasing a cheap case with a power supply already installed almost certainly means that the power supply is cheap. Such a purchase is best avoided. There is certainly no guarantee that a branded model will have

good EMC characteristics, but it does reduce the risk of EMC problems. Major manufacturers have a great deal of reputation to lose, even with just a few complaints.

What can be more certain is that the case and power supply will have been built up to a quality, all of which will have been through a rigorous EMC assessment routine.

As with any product, maintenance is important if the best performance is critical, and this goes with computers and peripherals. Ensure all screws are kept tight on covers, especially if they have previously been removed. Make sure that the finger-stock which ensures RF shielding is clean and in good contact. Finger tightening may not have the desired effect, as shown in **Fig 13.4.**

Make sure that screws on D-type connectors are tight. It is surprising how often these screws are not properly tightened by interface board manufacturers, and they only hold in place by friction or because

somebody has applied a dab of Loctite. They often unscrew when a cable is removed.

All screws should be tight and if they help to perform an earth connection you should ensure that there is no paint or other insulator under the parts.

Don't trust the tightening of screws, closing of gaps and installation of clip-on ferrites - check them yourself!

Laptop/Portable

There are some advantages for the use of laptop computers in the radio shack, and especially on the field portable site. Size is sometimes an important consideration, where the shack is in a very confined space.

From the point of view of EMC, laptops perform quite well. Because of the compact nature, radiation from the computer unit is quite well controlled, and this speaks well for the direct ingress of RF.

However, all modern laptops have an external, in-cable, power supply, and these are a potential source of RFI. Even the more expensive laptops use PSU's 'Made in China', and whilst some of these are built to a very tightly controlled specification, there is evidence that others are not.

Unfortunately, this is one of the areas where reading the specification is not a good indicator of EMC performance that will be good enough in the radio shack.

Possibly the biggest disadvantage of using a laptop is that the I/O ports may be quite constrictive. Many now have only USB and/or Firewire ports, which means that all level conversion for RS232 or TTL must be achieved outside of the unit. Unfortunately these products are not well screened, often being housed in plastic cases. It is important to examine the quality of items like USB/serial convertors, and USB port extenders. Look for products in metal cases!

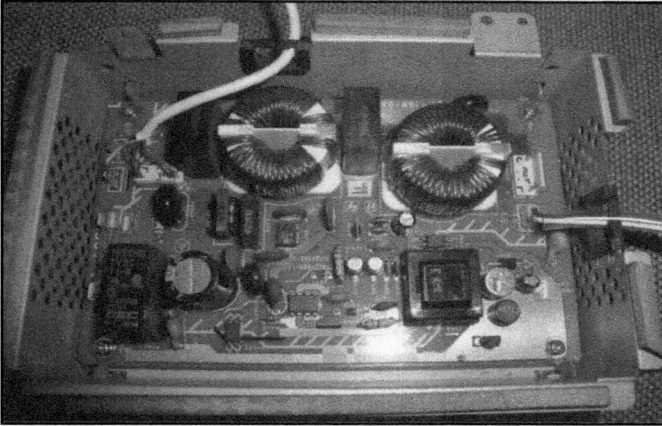

Fig 13.5: A power line filter recovered from a large-screen plasma TV. The filter components are across the top half of the PCB.

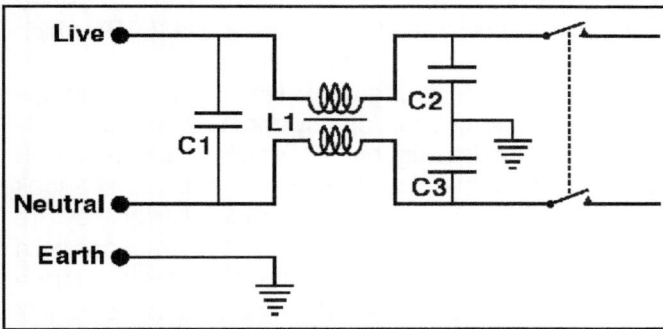

Fig 13.6: Circuit of a typical common-mode supply-line filter.

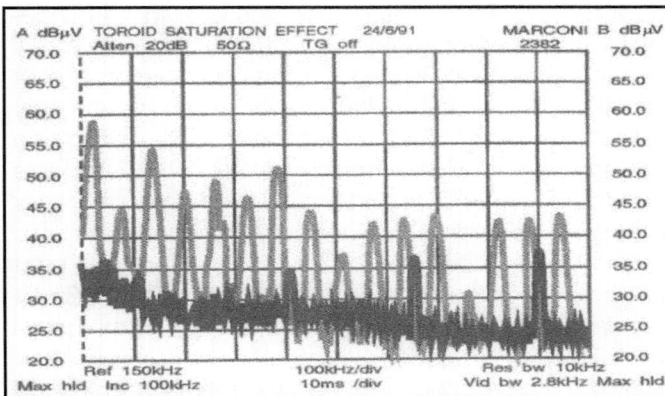

Fig 13.7: Performance of a typical common-mode supply-line filter. The plot shows the effect of saturation caused by not having the windings in counter-phase (in grey).

Mitigation Measures

However good the filtering, it should be remembered that it was provided at a price and the manufacturer will have been cost-conscious. Some additional mains filtering could prove valuable, both by providing a reduction in the leakage signals, but also by breaking the common mode impedance between the computer and its associated hardware, and the elements of the radio station itself.

A typical common mode filter looks like a low-pass filter, usually with two inductive elements, which are common mode chokes, and three stages of capacitive decoupling, perhaps using Y- capacitors. A very good source of these materials is old computer power supplies, or if you find somebody dismantling a defunct large screen plasma TV they have excellent filtering on the mains. **Fig 13.5** shows a unit recovered from an old Sony plasma, which also contains low voltage controlled power on/off switching, and as can be seen a good measure of additional screening. The circuit diagram of a typical common-mode supply-line filter can be seen in **Fig 13.6**, with its performance depicted in **Fig 13.7**.

The common-mode choke (L1) is wound with the live and neutral in counter phase on the same toroid, so that the currents cancel but the winding still provides effective in-line (common) impedance.

Whole Station Filtering

The advent of digital processing and more and more electronic devices in the home has resulted in what can best be described as 'radio fog'. Whilst the measures set-out in this chapter apply particularly to computers in the shack, it is worth remembering that a lot of 'rubbish' may be arriving into the shack via the mains supply. To this end the installation of a filter for the whole shack; indeed even the whole household, may be beneficial.

For the shack, which may consume up to 2kW, the installation of a filter is quite straightforward. A filter such as the one recovered from the plasma television will be quite suitable for loads up to 1.5kW. There are just one or two safety issues to remember - the capacitors and inductors must be suitable for purpose. The capacitors in particular must be X2 type across the supply and suitable Y types from Live and Neutral to ground. The inductors must have good insulation resistance and the whole assembly should be flash tested with a Megger before use. Your local electrical installer may be able to help do this.

Motherboards/Processors

Higher clock frequencies present less of a problem on HF, but are potentially more noticeable on VHF. However, most motherboards and I/O boards, have clock dividers/multipliers which result in a myriad of frequencies that could result in 'birdies' within the usable radio spectrum.

Most motherboards are very well manufactured from the point of view of RFI. They are built on multi-layer PCBs, with the outer planes often being ground or power supply rails. All of the active tracks are buried well away, inside the board. This is done, not only to control RFI, but more particularly so that the impedance of the tracks can be closely controlled. Many of the signal paths will be carrying bit rates well up into the 1GB/s region, so impedance control becomes critical.

All motherboards will have an array of clock oscillators and multipliers/dividers. It is therefore inevitable that some 'birdies' will be heard somewhere within the radio spectrum. The choice of a motherboard will come down to preference of a particular chip-set (Intel/AMD) and the user interfaces available, although these are becoming more standardised.

Cables and Interfaces

The interconnection of a motherboard and its many peripherals is important from the point of view of RFI. The quality of these cables has a significant bearing on the leakage of radiation. Better quality cables and the associated connectors will be less likely to cause a problem. Look particularly at the grounding of a plug into a socket, especially if it uses screws. Make sure that these fit well and are of course properly tightened/locked. Not only will this reduce the possibility of egress of interfering signal, it may also have an impact on ingress of RF into the computer.

In recent times there has been a tendency to migrate to serial interfaces for disk drives (SATA), which provides some advantage. SATA cables are run in a small screened bundle of uniform twisted pairs (UTP), which reduces/controls radiation.

Wherever a cable leaves a computer it is important to stop any common-mode signals from leaving the unit. This is usually achieved by clip-on ferrite chokes. Whilst these look ugly, the bigger they are the better they are at stopping the egress of RFI. In choosing a computer, make sure that all cables leaving the main computer unit, and at the input/output of all peripherals are fitted with ferrite chokes. If you are building your own machine, be liberal with 'common-mode' stoppers.

The tendency towards USB for all I/O devices connected to modern computers has led to the need for convertors, for example for serial or parallel devices. There are still very many of these around, especially in the radio shack. Many transceivers and TNC modems use RS232, and a number of antenna rotator controllers use parallel port connection.

If possible, these convertors are best housed within the tower cabinet. The USB port connections can be found on the motherboard and routed to the convertor, and its serial/parallel output routed to the rear panel. A good idea is to find the finger-plates from old unused interface cards - these will have standard RS-232 connector or parallel connector cut-outs and can be adapted to carry the connections to the outside world. Often these will have fully screen connectors with built in capacitors, so do not throw away the original connectors. It is a truism that some of yesterday's products were better built than today's!

A big advantage of putting all the hardware inside the computer case is that it is all connected to a common earth, thereby reducing the possibility of ground loops.

Of course it is still possible to purchase motherboards with at least two serial ports and there are suitable PCI - I/O cards which can add more serial and/or parallel ports. These, however, are not the norm.

Monitors

There is no doubt that the monitor of choice for the radio shack are flat-screen solid-state LCD models. Perhaps this will change in the future, with the rapid introduction ontp the market of Organic LED (OLED) monitors.

Very few RFI problems have been identified with LED monitors, but once again you should look to the interface cable as a potential source. Make sure that the screws are properly tightened and keep the cable away from live RF circuits.

Many monitors are independently supplied by small, Switched Mode power units. These are notoriously bad sources of RFI. The manufacturer may have fitted clip-on or moulded-on ferrites, but these are often insufficient to stop radiation at higher frequencies. It is advisable to fit additional ferrites on the power cord, and on the output cable of the supply. Trying another PSU, of suitable voltage and current rating may also help - quite often the switching frequency will be different, and a spurious signal may be moved away/outside of frequencies of interest. You may even find an old linear power supply of suitable size and ratings.

There is also the issue of various clock oscillators within the Monitor. The wise and additional precaution here is to fit clip-on errites to both the signal cable coming from the computer, place this as close as possible to the connector on the monitor, and once again

on the cable leading to the PSU. Finally, do not leave cables sprawled around behind the monitor/computer. It is best to coil them up, not only to make them look neat and tidy, but more particularly to minimise radiation and pick-up of RF in the shack.

External Hardware

Already mentioned is the issue of common-mode signals and how they can be prevented, but it is worth emphasising the need to connect all parts of the system to one common earth point and ensure that any cables do not carry common mode signals by the generous use of clip-on common-mode ferrite chokes.

Connections to/from transceiver(s) require special attention. Most CAT interfaces use the RS232 interface standard, although some (notably Icom) have adopted a TTL interface.

Whichever is used, it is good practice to use an optically isolated interface on data and control lines. This may either be built within a purchased interface adapter or with a home-brew adapter built to one of the many published designs.

TNCs and issues connected with these are somewhat similar to the CAT interface, except that the modulation signals need to be considered. In practice, by far the best way of dealing with low frequency audio circuits is by the use of 1:1 isolation transformers. Although these are getting a little harder to source, they can be found in older surplus modem interface cards.

The more expensive solution for those who do not have an aptitude for construction is one of the ready-made multi-mode interface units, which handle keying, data-mode modulation schemes as well as the essentials of the CAT interface.

For the homebrew constructor wishing to deal with sound card input/outputs, there are a number of published circuit configurations, the simplest involving small audio transformers to achieve physical isolation, and thereby ground loop problems which may result in RFI/ EMC problems.

When tracking-down RFI it is important not to forget direct radiation from interface devices. TNCs, multimode adapters, bridges/ routers and the like that will contain micro-processors clocked at frequencies up to or even greater than 24MHz. Once again these may be divided or multiplied, thus providing a spectrum full of birdies.

When investigating these problems, start with everything turned off except for the station receiver. Turn things on one at a time, investigating at each stage what unwanted signals have appeared. You should certainly find some signals and you will have to assess whether they warrant tracking down.

Most computer clocks can be identified. Being crystal oscillators thay are reasonably stable, if not a little raucous in tone. However, in recent years there has been a tendency towards the use of 'dithered' clock oscillators. These are quite difficult to identify and sound more like a noise source. The only way of being certain that they are a source of inter-ference is to turn them off.

The technique of dithering a clock os-cillator spreads the spectrum created, thus reducing the Power Spectral Density. Since EMC measurements are made with defined bandwidth filters, spreading the signals across the spectrum reduces the peak level, making it easier for a manufacturer to meet the limits in the standard.

Positioning of Equipment and Cables

The concept of a single earth, both to avoid ground loops and also RFI/ EMC problems, is difficult within the entirety of a radio shack. However, every effort must be made to achieve the very minimum number of earth connections between the computer and as-sociate peripherals, and the radio equipment proper. If connections have to be made it is advisable to break-up the common imped-ance with the ubiquitous CM choking -a clip-on ferrite or several turns through a ferrite ring.

Separating data cables from RF signal circuits is an important objective, which might mean positioning the computer and all pe-ripherals at one end of the shack bench and taking all live RF circuits to the other end. This requires some careful planning, but separa-tion may pay dividends in the long run. Most importantly it reduces the possibility of mag-netic field coupling between the computer system and the radio.

Whilst it may be easier to deal with near field magnetic coupling by moving things further apart, the real issue which will confront radio amateurs, most of the time, will be the far-field electric field, which does not decay quickly with separation distance.

Many radio amateurs will wish to be con-nected to the Internet when operating, for the many reasons given elsewhere. Best practice suggests that a wireless system is safest in the shack.

It avoids the issues of coupling of RF along networking cables, although many prefer this type of system because of the security it af-fords. When RF is coupled into UTP wiring it inevitably gets into the home hub or router and seriously reduces data rates and may cause the connection to the remote server to drop.

It almost goes without saying, but must be said - whatever you use, avoid PLT or any similar power line communications products at all cost.

Finally, when you think you have paid at-tention to all of the potential problems, there will be others! However, most problems have a solution. It may take a little experimenting, but that is what rmateur radio is all about.

Telephones

In the days of landline telephones with me-chanical dials, they contained little in the way of electronics and were not particularly susceptible to interference from nearby transmitters. These days its a different story, because practically all landline telephones on the market contain electronics. Cordless domestic telephones invariably contain a lot of electronics.

It is the active, amplifying elements of modern telephones - corded or cordless -

ADSL

At the present time, the most popular way of accessing the Internet from home is by ADSL, which uses frequencies up to about 1MHz, these signals being passed along a standard 2-wire telephone line. Amateur radio transmissions of 1.8MHz and up in frequency should not affect it, but unfortunately it is not uncommon to hear of ADSL lock ups occurring when transmissions are made on the 1.8 or 3.5MHz bands. Naturally it depends on how close the transmitting antenna is to the telephone wiring and how good the telephone wiring is, but telephone wiring and how good the telephone wiring is, but sometimes it only requires a few watts of RF to cause a problem.

Plug-in ADSL line filters are good at what they do, but as already stated they do not filter the ADSL signal. Enter a second problem. A lot of domestic telephone wiring is based on a wiring plan which inherently unbalances what should be a balanced circuit - the telephone line itself. This is because a lot of master sockets contain circuitry to extract the ringing signal from the incoming line and send it along a third wire within the home. Although the telephone line is now unbalanced, it does not usually affect ADSL performance adversely... until RF is radiated nearby. It is also worth noting that ADSL line filters do not pass the 'ringer' signal through, because modern telephones don't need it.

As standard, the incoming telephone line will be connected through to pins 2 and 5 of the telephone socket, with the circuitry in the master socket extracting the ringer signal and passing it through to pin 3. If you can extract the ADSL signal before through to pin 3. If you can extract the ADSL signal before the master socket, any unbalanced part of the system will be 'downstream' and should no longer cause a problem.

If your master socket looks like the one shown in

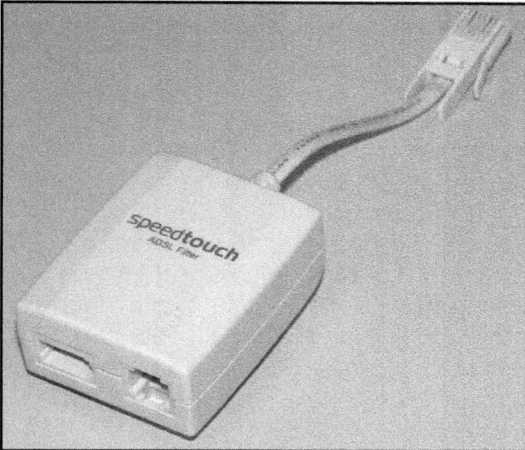

Fig 13.8: An ADSL line filter, which can be used to filter the line for a telephone which is suffering RF breakthrough.

that lead to interference being caused so readily to them, but fortunately there are a number of steps that can be taken to reduce or eliminate it.

Today, probably the easiest and simplest thing to try and remedy a landline telephone that is suffering breakthrough is a line filter used for ADSL (Asymmetric Digital Subscriber Line). ADSL filters are readily available and plug directly into telephone sockets. A Speedtouch ADSL filter is shown in **Fig 13.8**. Other makes and models are also available.

An ADSL filter will have two sockets on it - one for a telephone and the second for an ADSL modem. As **Fig 13.9** shows, the circuitry inside an ADSL line filter consists of a low pass filter to remove the higher frequencies used by the ADSL signal from the telephone socket. The signal from the incoming line passes straight through to the ADSL output socket, with no filtering.

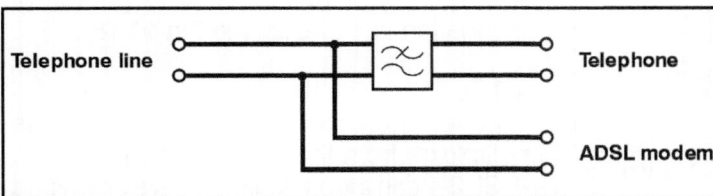

Fig 13.9: Internal arrangements of an ADSL line filter.

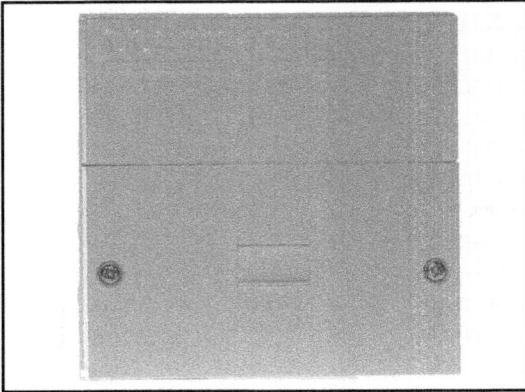

Fig 13.10: Old style telephone socket. Master and extension sockets look the same externally, but there are three extra components inside a master socket.

Fig 13.11: Standard BT type NTE5 master socket.

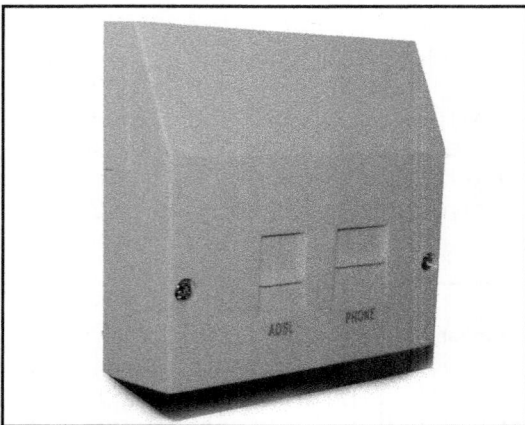

Fig 13.12: NTE5 master socket with re- placement faceplate that has a built-in ADSI line filter.

Fig 13.10 it is the old style and your best option is to plug an ADSL line filter into it and position your ADSL modem close to it. If you can gain access to the telephone line where it enters the house, it could be worthwhile adding a clip-on ferrite ahead of the master socket.

If your master socket looks like the one shown in **Fig 13.11** (note the fact that the screws are near the bottom and there is a dividing line just over half way up) it is a new style master socket (NTE5) and more possibilities for interference reduction exist.

Whilst you are still not entitled to make any modifications to the line or the master socket itself, the master socket is in fact hidden inside and it is possible to replace the faceplate with one that has an ADSL line filter built into it. This results in a master socket that looks like the one shown in **Fig 13.12**.

This type of arrangement has two distinct advantages:

1. The filter for all the telephones is built into the master socket, so there is no need to use an ADSL line filter for any of the individual telephones.
2. As **Fig 13.13** shows, the ADSL signal also appears on the connections inside thefaceplate, so if you do not wish to locate your ADSL modem near the master socket you can add a twsired pair to conduct the signalaway to where you want it.

If required, further filtering can be added to the separated ADSL signal, by fitting a suitable low pass filter. The design shown in **Fig 13.14** is attributed to OZ7C. Standard value components can be placed in series and/ or parallel, to achieve the required values. It is stated as having a 1dB ripple uo to its cut off frequency of 1Mhz and offering 40dB of attenuation at 1.8MHz (see **Fig 13.15**) .

ADSL2

Depending on the distance of the access multiplexer from the subscriber's premises, ADSL 2+ can theoretically achieve download speeds of 24Mbit/sec and upload speeds here is that the 'good practice' BT adheres to should be maintained right throughout the wiring. By choice, the modem will be as close as practical to the master socket. All in-house wiring beyond the modem should at the very least comply with the CAT5 or - even better - CAT6 standard. RS Components and other online suppliers stock suitable cable with 5 Uniform Twisted Pairs (UTP), and connector boxes.

The filter shown in Fig 13.14 is not suitable for homes served with ADSL2, since it cuts-off at too low a frequency. For ADSL2 frequencies up to 2.2MHz are employed.

If you have issues the best you can do for yourself is fit a Service Specific Faceplate (SSFP) fitted to eliminate the internal wiring, since

(a) internal wiring acts as a significant antenna system to pick up interfering signals, and (b) the SSPF will eliminate the additional unbalance that this 3-wire network can create. If a SSFP is not suitable because you require the flexibility provided by plug-in micro-filters, the BT Broadband Accelerator plate (www.bt.com/accelerator) may help improve things, although this is only suitable for ADSL and ADSL2+. This however does not matter since BT Infinity is currently installed with a SSFP only. BT also has a common mode filter, the BT80A-RF3 (an example can be seen on www.kitz.co.uk/adsl/ btsockets. htm), that engineers will fit if they suspect noise issues (for example from SMPSUs), but I don't know how you would go about

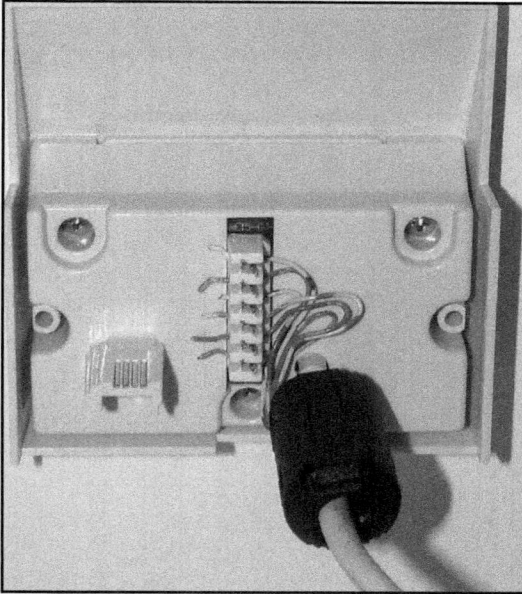

Fig 13.13: Inside a faceplate that has a built-in ADSL line filter. Extension telephones and a remote ADSL modem are connected via the cable. Plugs into master socket connections to phone extensions and ADSL modem clip-on ferrite.

of3.5Mbit/sec. Using the trading name BT Infinity, BT is actively rolling it out across the UK. Early experience is good, although there have been reports of interference with 160m (Top Band), causing broadband to drop out. The simple cure is to get the modem to train-up in the presence of the amateur transmission. Be aware that if the modem retrains at another time (for example after it has been power cycled), it will need to be trained again in the presence of the amateur signal.

Instances of interference have been rare, which is as a result of the very good balance of the network cables and wiring, right up to the modem. BT do not use stub connections within their network, which might otherwise cause imbalance and radiation. An important point to remember

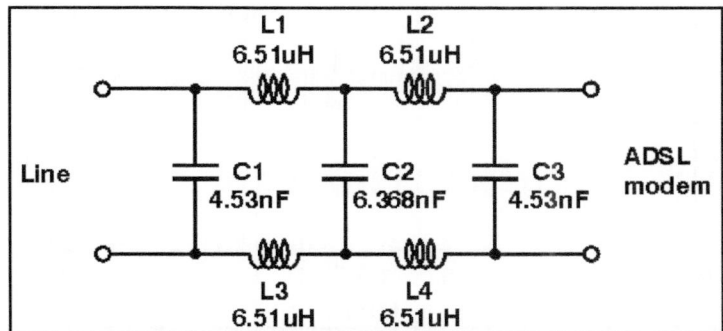

Fig 13.14: ADSL filter attributed to OZ7C.

requesting this item from BT.

Sites like Kitz (www.kitz.co.uk) provide a lot of very useful infor-mation on broadband and how it works, issues etc, with the section at www.kitz.co.uk/adsl/rein.htm dealing with interference from faulty power supplies (what BT call REIN).

If you suspect your broadband service is being interfered with by REIN (it could be a regular pattern every evening or at a particular time of day and fine at all other times), try listening on a battery powered Medium Wave radio tuned off station when you are experi-encing broadband problems. You may hear a horrible buzzing noise, especially if the radio is placed near the telecom network cable. If you do hear this, turn off your mains power. If the noise disappears the source is almost certainly within your own home and you should trace it by turning circuit breakers off one at a time and isolating individual items of equipment. (Note that faulty power supplies can also be incredibly hot to the touch!). If you turn all the power off and the buzzing

does not disappear the problem is likely to be within another property in the neighbourhood, in which case you could talk to your neighbours to see if they have similar issues at the same times and perhaps get them to do the same power down of their property.

Finally, if there is excessive RF around the shack (or household), the use of clip-on ferrites may eliminate RFI on the phone connections, especially if the in-house network is old and runs alongside mains cabling.

Fibre

Fibre broadband, carried over optical fibre cables, can offer speeds of 35-60Mbps and subject to you location and the infrastructure available as much as 300-900Mbps. Depending on where the fibre terminates, the cause and effect of interference are likely to be the same as ADSL2 but with less potential for remote external interference.

4G & 5G

These mobile phone technologies also offer useable data transfer rates. 4G is typically 8-10Mbps while 5G is potentially much faster at up to 20Gbps (Giga bits per second).

Knowledge and experience of interference from mobile is widespread. Their use in or near the radio shack should be avoided and, where used for internet connectivity, they should be installed as far from the shack as possible. Thereafter cable-bound interference can be managed in the same way as ADSl cabling etc.

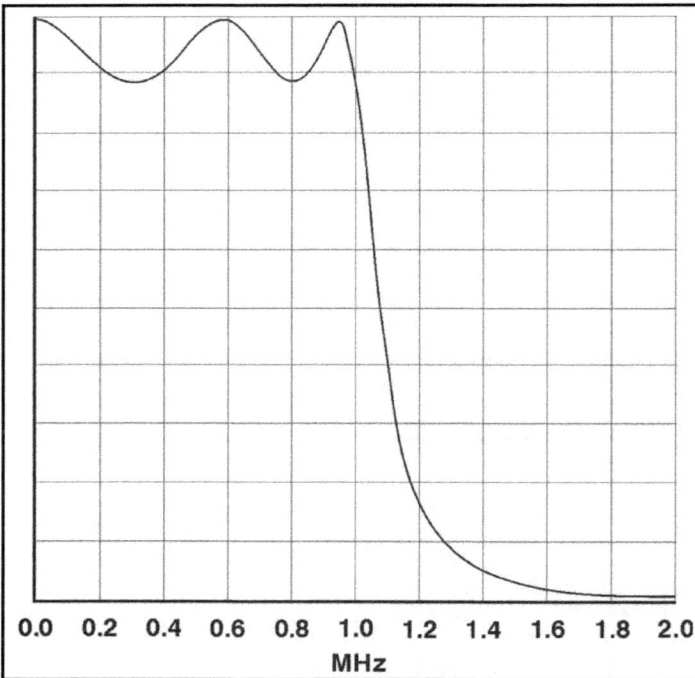

Fig 13.15: Performance of OZ7C's ADSL

14
Internet Linking

by Lorna Smart, 2E0POI
(with information from another source)

These days there are several ways in which radio amateurs can communicate without the straightforward use of a transceiver. Some Internet-linked systems don't involve radio at all. There has certainly been some developments in this area since this book was last published. Some more established ways of internet linking are:

IRLP
Echolink
eQSO
WIRES-II
CQ100

One of the newer methods of internet linking is Zumspot Rpi Digital Voice. Below is a review written by Mike Richards, G4WNC that was published in the March 2021 edition of RadCom Plus.

How do you decide whether to follow the DMR, D-Star or Yaesu Fusion route for your handheld? You may not need to decide.

The 2m and 70cm bands used to be the lifeblood of local amateur radio clubs and the primary source of the local chatter. Over recent years, the previously analogue systems have been largely superseded by digital voice systems that offer many benefits, such as being able to use 70cm to communicate with fellow amateurs all over the world. However, new systems often mean new problems and, in the case of digital voice, it is the introduction of three similar but incompatible encoding systems, ie DMR, Yaesu Fusion and D-Star. Whilst DMR might seem the most logical way forward due to the commercial take-up and consequent opportunities for using surplus kit, all three systems have strong user bas-

es, so are likely to remain with us. As you might expect, a few enthusiastic amateurs have tackled the compatibility problem, and one of the solutions is to use what's become known as a hotspot.

What is a hotspot

If you're a fan of the Douglas Adams' Hitch-hikers Guide to the Galaxy, you will have heard of the Babel Fish. It's a small, bright yellow fish that, when inserted in the ear, allows the user to understand any language. A hotspot is somewhat like that Babel Fish as it enables working other digital voice modes with your single-mode handheld. For example, as shown in Photo 41. with the ZumSpot, you can use your DMR rig to talk to a fellow amateur using a Yaesu Fusion rig. Impressive as that may seem, a hotspot takes the technology a bit further and uses its internet connectivity to route your call to

Photo 14.1 The ZUMSpot and handheld

any other network-connected repeater. That same link can also be used to connect to multi-mode servers that carry out the mode translation for you. This enables contacts with a wider range of digital modes such as D-Star, P25 and NXDN. The hotspot can therefore be thought of as your own private repeater that can be used for local calls or to link farther afield via the internet.

There are two main components to the ZUMSpot RPi; a single board computer (SBC) and a radio board. In the ZUMSpot RPi, the SBC is a Raspberry Pi Zero W, which is the Wi-Fi and Bluetooth enabled version of the single-core Pi Zero.

The radio board uses the same physical profile as the Pi and stacks neatly on the Pi's GPIO pins. Powered by an STM32 ARM microcontroller and running MMDVM firmware, the radio board provides the mode translation service and drives an ADF7021 RF transceiver chip to handle the RF link to your transceiver. The ZUMSpot also includes a small 1.3" OLED display that piggy-backs on the radio board and displays status information for the hotspot, see **Photo 2**.

The Pi-Star software running on the Raspberry Pi provides the user interface for configuring the ZUMSpot RPi. This software is extremely powerful and versatile and can be used to power a full-blown repeater.

Getting started

The ZUMSpot from Martin Lynch and Sons comes with all the software and firmware pre-installed, which is very helpful. However, I still had to run through some basic configuration steps in Pi-Star to add my station specific details. Although the ZUMSpot RPi is supplied with very little paperwork, there was a link to the excellent Pi-Star guide by Toshen, KE0FHS. This guide provided a very clear path through the configuration steps with plenty of illustrations to help along the way. The first step was to get the ZUMSpot RPi connected to my home Wi-Fi network. This turned out to be a simple operation thanks to Pi-Star's Auto AP(Access Point). If, during startup, Pi-Star fails to find a network connection, it will automatically activate its local Wi-Fi AP. I was able to login to this AP using my smartphone and then follow the menus to enter the SSID (Service Set IDentifier) and password for my Home Wi-Fi network. Once the ZUMSpot RPi was rebooted, it found and connected to my local network. With the ZUMSpot RPi up and running, I could connect to the main Pi-Star interface simply by entering pi-star.local as the URL in my browser. This took me straight to the Pi-Star Dashboard, **Fig 14.1**, that provided a helpful status summary for the ZUMSpot RPi as well as giving access to the configuration menus.

Before I could start operating, I had to add a few personal details in Pi-Star, such as my name, callsign, DMR ID, location, etc. As you would expect, the Pi-Star configuration menus are password protected and have a default username/password combination. One important point here is that you must change the default password. As the ZUMSpot RPi is connected to the internet, running with the default password is a security risk. Once you've made the change to a strong password, the ZUMSpot RPi becomes very secure.

Once I had access to the configuration menu, the first task was to

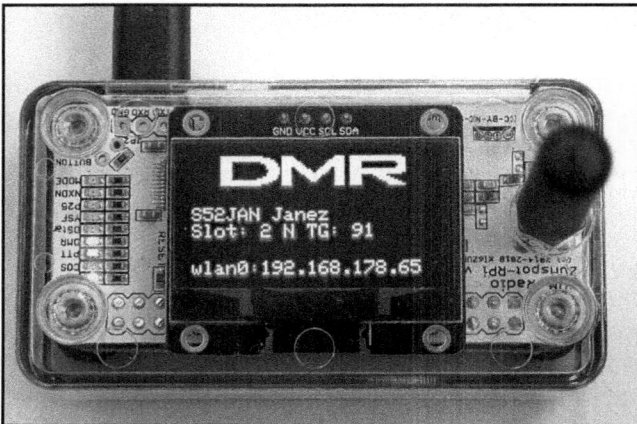

Photo 14.2 ZUMSpot status screen

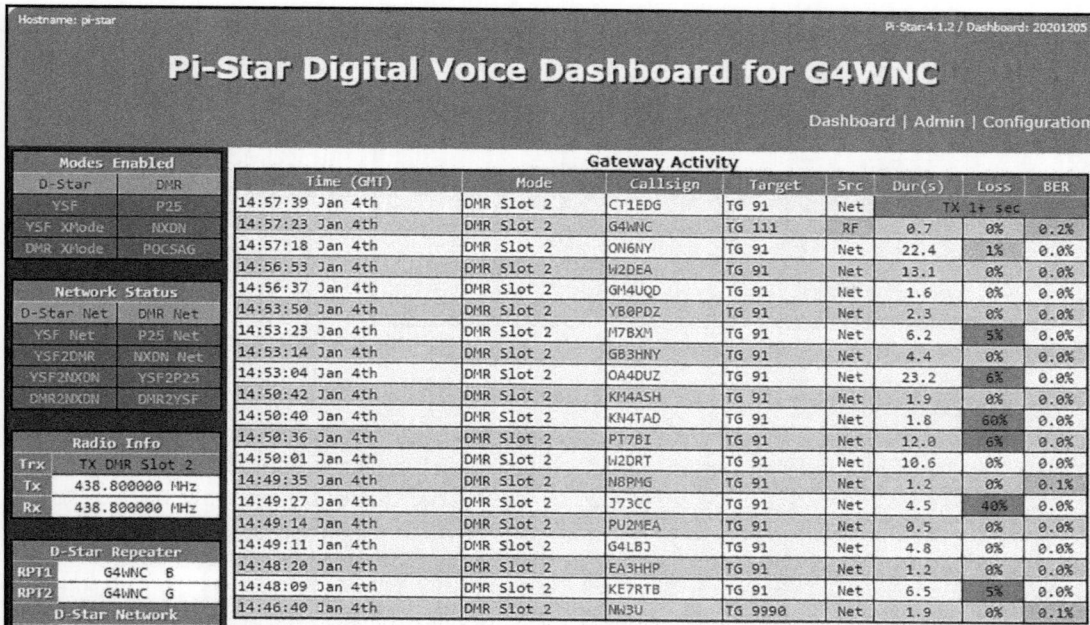

Fig 14.1 PiStar dashboard

enable the appropriate digital radio modes in the MMDVM Host Configuration section, **Fig 14.2**. Pi-Star includes all the popular digital voice modes such as DMR, D-Star, Yaesu Fusion, P25 and NXDN plus conversions between some modes. For most users, it's generally best to set Pi-Star to match your handheld, which in my case, was DMR. I say this because the online cross-mode servers do a very good job of mode conversion so you get a slicker changeover between overs than when working cross-mode locally. If you

are planning to do cross-mode scanning on the ZUMSpot RPi, you may need to adjust the hang time settings. The hang time is the length of time Pi-Star will remain in the current digital mode after a transmission ceases. It's important to get this right because, if it's too short, Pi-Star will revert to mode scanning between overs. Pi-Star then requires around 1.5 seconds of audio to identify the mode before continuing the QSO. When the hang time is correctly set, Pi-Star should remain in the selected mode in between overs, thus giving a smoother exchange and making it easier to follow QSOs.

On completion of this section, I had to click the Apply Changes button to save the changes and add the DMR configuration panel. This also applies to all the other modes, as you won't see the relevant configuration screen until you've set the mode and applied the changes.

Another step that may be necessary is to adjust the Tx and Rx offset. This is used to compensate for small frequency errors in the reference clock that drives the transceiver chip. Any error here shows up as an in-

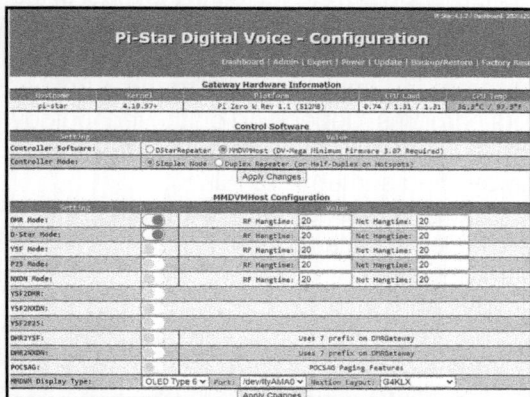

Fig 14.2 MMDVM Host configuration

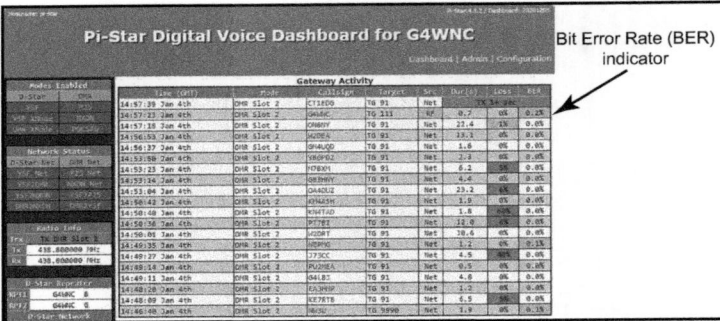

Bit Error Rate (BER) indicator

Fig 14.3 BER display on the dashboard

creased bit error rate (BER) that you can see reported in the Pi-Star dashboard, Fig 14.3. The review model was spot-on frequency and returned a BER of 0.1% or better, which is good for a DMR link. I experimented with the offsets to see if I could make any further improvements. To access the clock offsets, I opened the Configuration menu and chose Expert followed by MMDVMHost and then altered the RXOffset and TXOffset in the Modem section. I tried positive and negative offset values and found a useful starting point was to adjust both offsets in 100Hz steps. After each adjustment, I hit the Apply Changes button and returned to the dashboard. I could get an updated BER reading by keying the handheld and checking the BER column of the Gateway Activity monitor. I found this to be an effective way to adjust the offsets.

Next I proceeded to complete the DMR configuration by selecting the BrandMiester network.

Using the ZUMSpot

For most users, the simplest way to use the ZUMSpot RPi is to configure it to match the digital mode of your handheld, ie DMR, D-Star or Yaesu Fusion. You will still be able to work crossmode, but the mode conversion will be done by the online servers rather than in the ZUMSpot RPi. This generally allows for a faster exchange of overs and greatly simplifies setup. Before I could use the ZUMSpot RPi I had to configure my radio with the operating frequency and the talkgroups, reflectors, etc that I wanted to use. As I was using a DMR rig, I started by creating a new Zone for the ZUMSpot RPi and then added channels for each of the popular talkgroups such as Worldwide (91), Europe, UK, etc. It's also important to remember to set your rig to low power for all the programmed channels as you won't ever need to use a higher power with a hotspot. From this point on, using the hotspot was extremely simple. I just selected the talkgroup on the rig, keyed up for about a second and the ZUMSpot RPi would switch to monitoring that talkgroup and I could listen on my rig. To make a call I simply keyed up on the desired Talkgroup. At this point I was free to roam the house and garden whilst in QSO. I tested the range with the AnyTone AT-D868UV set to its low power setting of 1 watt. I was able to operate successfully anywhere in the house and also at the extremities of the property, which were about 35 metres from the ZUMspot Rpi.

Summary

The ZUMSpot worked very well during the review and transformed the capabilities of a handheld rig. It literally provides a gateway to the world from any location that has Internet connectivity and is an ideal partner for most digital handheld rigs. To make the most of the ZUMspot Rpi, you do need to have a good working knowledge of your rig, so you can program the appropriate talkgroups, reflectors or rooms, depending on the technology you're using. With ZUMspot RPi, I could always find someone to talk to.

The ZUMSpot RPi costs £159.95 and is available from Martin Lynch and Sons (www.hamradio.co.uk).

My thanks to Martin Lynch and Sons for the loan of the ZUMSpot, AnyTone AT-D868UV (£129.95) and TYT MD-UV380 (£84.95) DMR radios

15
Interfacing and Interfaces

by Steve White, G3ZVW, revised by Lorna Smart, 2E0PO

There are numerous commercially-made items of equipment that will interface between a computer and a radio. Some are complex and some are so simple that those with basic knowledge of electronics should be able to build them. In this chapter you will find information on commercial and make-at-home interfaces.

Build Yourself

Non-isolated
The simplest circuit that can be used to connect a computer with an RS232 serial port (or a Centronics parallel port) to a transmitter requires just one resistor and one general-purpose NPN transistor. It is suitable for keying CW and PTT. For use in keying the PTT, make sure that your transmitter has diodes across the changeover relays to suppress the back EMF that inevitably arises when they release, or the transistor may be destroyed the first time the PTT signal goes off!

The circuit is shown in **Fig 15.1**. Typical connections are shown in **Table15.1**. Such circuits are often built into the body of a D-type connector.

A superior circuit is shown in **Fig 15.2**. This is more likely to be built into a small project box between the computer and transmitter.

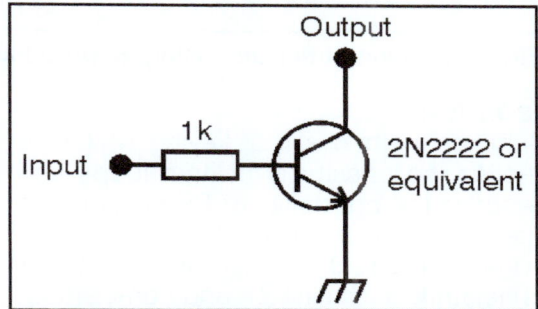

Fig 15.1: Simple CW keying and PTT interface.

Keying	9-pin RS232	25-pin RS232	Centronics
Input	4	20	17
Ground	5	7	1

PTT	9-pin RS232	25-pin RS232	
Input	7	4	
Ground	5	7	

Table 15.1: Simple CW keying and PTT interface connections.

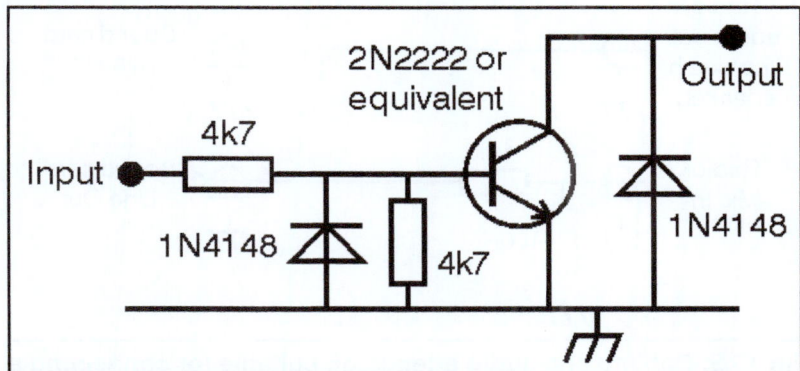

Fig 15.2:Superior non-isolated interface.

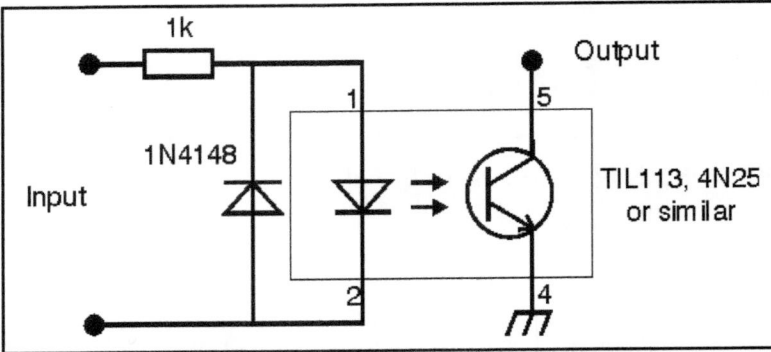

Fig 15.3: Keying interface using an opto-isolator.

Fig 15.4: The pinout of typical opto-isolators.

Isolated

A better method still of keying which removes the possibility of hum loops and reduces the possibility of RF pickup is to use an opto isolator. The circuit for this is shown in **Fig 15.3**. The type of opto isolator is not critical. Darlington and non-Darlington types should work equally well. The pinout of many opto isolators is as shown in **Fig 15.4**. If you use a different type to the ones mentioned in the circuit, check before you build the interface that the pinout is the same.

Once again, if you are going to use a circuit such as this to key the PTT, make sure the relays in the transmitter have diodes across them.

Audio

For datamode operation via a sound card, screened cables are called for. The simplest connection will be two screened cables, one to carry audio from the sound card of the computer to the radio and a second to carry audio from the radio to the sound card. If your transmitter has a 'phone patch' (external, high level audio) input, it is recommended that you use it in favour of the microphone input. This is because most transmitter microphone inputs are very sensitive and it is all too easy to overload the input stage, which will result in distorted audio when you transmit. If you connect to a high level audio socket, remember to unplug your microphone during datamode operation or it is possible that it will pick up and transmit the sound of whatever is going on in your shack while you are transmitting.

If your radio does not have a phone patch input you should build the circuit shown in **Fig 15.5**, which includes a 100:1 attenuator. If your radio does have a phone patch input, it is likely that you will be able to omit the two resistors and connect the sound card output direct to the radio input. Either way, adjust the microphone gain to a level that does not cause the audio stages of the transmitter to be over driven.

If you experience problems with RF

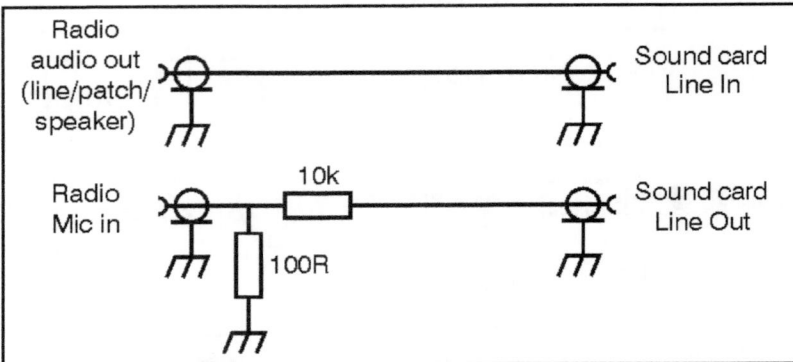

Fig 15.5: Cabling and audio attenuator, suitable for connecting a computer to a radio that does not have a high level audio input.

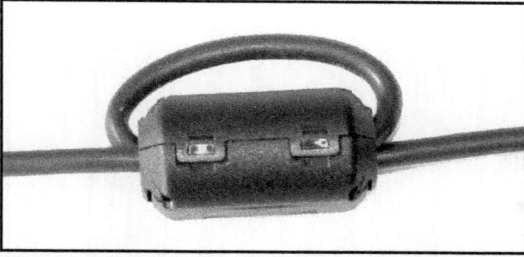

Fig 15.6: A clip-on ferrite correctly installed on an audio cable.

Frequency	Amidon type Number	Colour
LF	FT240-31	Dull black
LF-HF	FT240-43	Shiny black
HF	FT240-61	Dull black

Amidon type Number broken down

F =	240 =	61 =
Ferrite	Diameter (240 means 2.4 inches)	Material type

Table 15.2: Ferrite rings/cores suitable for winding on audio cables.

pickup when transmitting, the first remedy to try is clip-on ferrites on the cables. It is good practice to wind an audio cable through a ferrite several times, but make sure the two halves close firmly together when you clip it shut or it will have no effect. **Fig 15.6** shows a clip-on ferrite correctly installed. Alternatively, use ferrite rings of an appropriate material for the frequency that is causing problems. Unless you are starting with a cable on which connectors are not already fitted, select a core size through which the smallest of the connectors can easy pass, and wind several turns onto the core. The cable should not be wound randomly, rather it should be wound from a start point and and around no more than 3/4 of the ring. Secure each end with a cable tie. Alternatively, the cable should be wound around about 3/8 of the ring, then crossed over and wound around another 3/8 of the ring. This is more effective, as it reduces capacitive coupling between the points where the cable enters and exits the ring.

There are big differences between the materials used in differing varieties of ferrite ring, so not all types of material are suitable for all frequencies. Suitable types to use are shown in **Table 15.2**. Do not use iron dust cores (eg T200-6), which can often be identified by the fact that they are colour coded.

If you experience hum on your data transmission, caused by a hum loop, you may need to use audio transformers to electrically isolate the computer from the transmitter. The circuit for a suitable system is shown in **Fig 15.7**.

Audio transformers can be recovered from ancient transistor radios, PC modem cards, or purchased from various retailers.

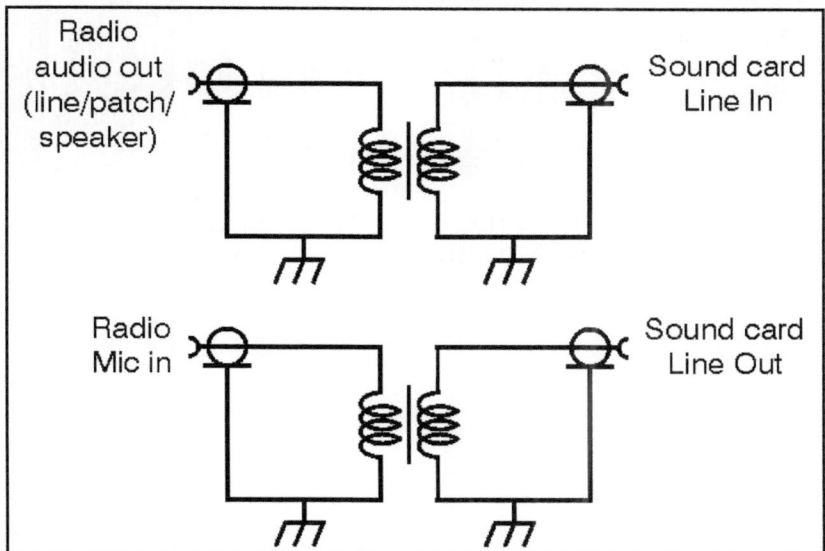

Fig 15.7: How audio transformers can be used to electrically isolate a radio from a computer.

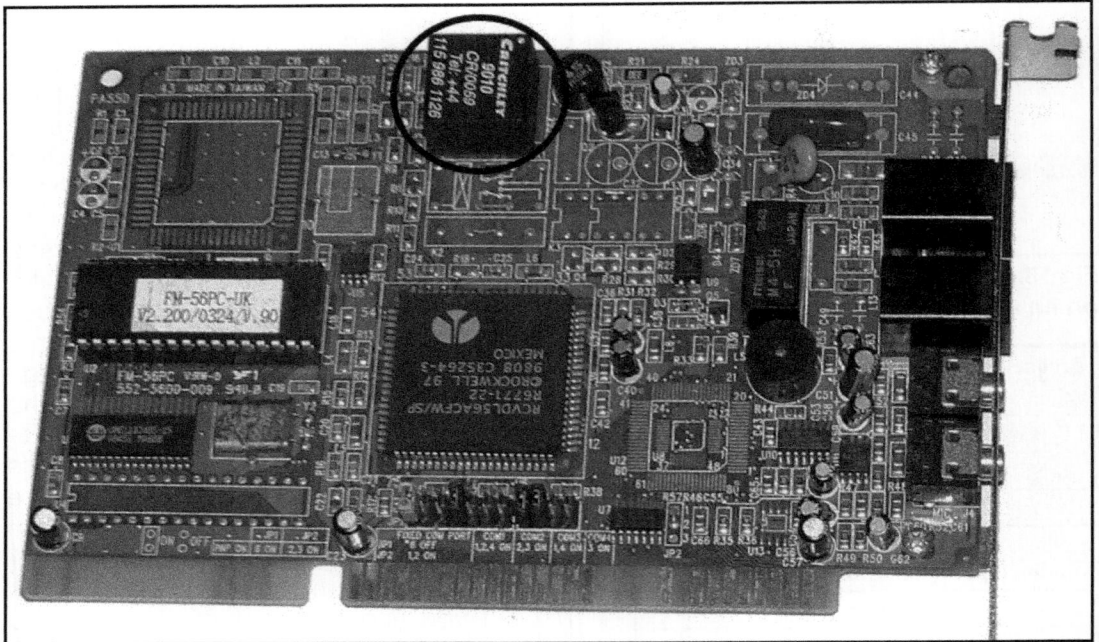

Fig 15.8: An old modem card from a computer, with the audio transformer circled.

Fig 15.8 shows an old modem card from a PC, with the audio transformer circled.

Kits

There are numerous CW keyers available as kits, but perhaps the most popular computer-linked model is the WinKeyer2 from K1EL (see **Fig 15.9**). It conforms to the WinKey standard and integrates well with many popular logging programs.

The kit is available in RS232 and USB versions and can take an experienced constructor less than two hours to build .A range of products for various purposes are available from a number of manufacturers.

Fig 15.9: The WinKeyer USB from K1EL.

16
Live Internet Applications

by Lorna Smart, 2E0PI

In the rapidly evolving technological world that we live in, applications come and go so quickly. Below, I have provided information and website addresses to some of the live internet applications that are available at present. The information provided comes from the previously published chapter of this book written by Steve White, G3ZVW which has been condensed.

Repeater Streaming Internet Feeds

There are a range of ways you can now to listen to repeaters when you are not near your radio. A lot of repeaters now provide their own internet steams. You can find streams from repeaters from all over the world at www.radioreference.com/. Zello www.zello.com/ , a mobile phone app also provides a directory of services that you can listen to. This is free of charge.

Chat Rooms

Chat rooms are not something you hear about so much anymore, however, they are still a viable means of communicating with other amateur radio users. ON4KST www.

on4kst.com/ provides chat rooms to users from all over the globe for no cost. Dxmaps. com www.dxmaps.com/ also has a plethora of chat rooms available. I suspect that there are chat rooms out there catering for all aspects of the hobby.

Online Receivers

These have evolved from the type were only one person could use them at at time. There are now Software Defined Receivers (SDRs) available online which can be used by more than one person at the same time. You can find a list of online SDRs at www.websdr.org.

Reverse Beacons

The Reverse Beacon Network (RBN) has come into being due to Software Defined Receiver (SDR) and Skimmer software. It is a worldwide network of SDRs whose purpose is to listen continuously to the amateur bands. They then report what stations they hear, the location and signal strength. To access this, you need to log into www.reversebeacon.net

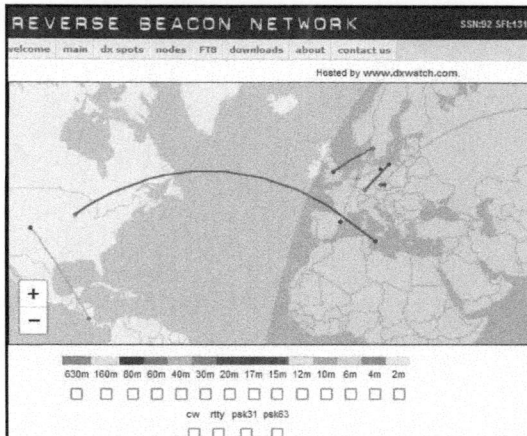

PSK Reporter

PSK Reporter allows the user to search by band, callsign, country, Locator square, mode and time. This will then produce a reports on whether a station is being received and the strength of it. The website is www.pskreporter.info.

DX Clustering

The DX Cluster has been in existence since the mid 1980s and is a near real-time alert system. There are a good selection of websites and applications for DX clustering. One such site is www.dxsummit.fi.

Live VHF DX Maps

These place the information from the DX Cluster network onto a map. This can be found on www.dxmaps.com/spots/map.php. They provide a way for VHF and UHF users to see where recent QSOs have happened.

Near Real-time Magnetometer

This was set up to provide advanced warning of possible auroral activity. It can be accessed via the website of Roger Blackwell, GM4PMK who set the Magnetometer up at www.marsport.org.uk/observatory/index.php.

Near Real-Time MUF Map

This is a map of Maximum Usable Frequencies(MUFs). It can be found on www.spacew.com/www/realtime.php.

17
The Raspberry Pi

by Mike Richards, G4WNC

Most of this chapter has been drawn from another RSGB book by Mike Richards, G4WNC called "Raspberry Pi Explained For Radio Amateurs" which is available from the RSGB directly https://www.rsgbshop.org or other online retailers.

1. Introduction

There are already plenty of Raspberry Pi books that cover the Pi from many angles, and I have bought many of them myself. However, I've yet to find one that feels like that vital reference you grab when you start a project. If you're anything like me, you are probably thinking 'what can I get the Pi to do right now'! So, I have started with a series of projects so that you can quickly make your Pi do something useful. This is followed by how to install popular radio applications and using the Pi for Software Defined Radio (SDR). After a while you'll want to create your own projects and that's where the other chapters will help you. At this point, I also ought to come clean and make you aware that I am not a formally trained programmer. Whilst I have trained as a communications engineer, most of my computing and much of my electronics skills have been self-taught, driven by a desire to build a variety of projects. These ranged from communications receivers in my youth through to PA systems and music electronics as I joined the obligatory 60s band. That was followed by a wide range of amateur radio related projects along with an assortment of photographic applications. I was fortunate in living through the birth of the home computer and being able to learn using the likes of the Ohio Superboard, BBC-B Micro and the Amiga computers. One of the most frustrating aspects of this type of project-driven learning is the speed with which you forget the, rapidly assimilated, knowledge. My personal solution to that problem has been to keep a blog, where I record the important or hard-found learning points, so I can quickly familiarise myself when I return to a similar project at some later date (see www. g4wnc. com). This has proved very successful for me and saves a lot of time and effort. I therefore decided that a similar approach might work well for this.

Like it or loath it, the Raspberry Pi generally works at its best when running the Raspbian distribution of the Linux operating system. This distribution is continually evolving and provides easy access to a wide range of programming languages. An additional, and important, benefit comes from the huge selection of code libraries and hardware support that is built into the distribution. The net result is that Raspbian is far and away the most popular operating system for the Pi which in turn, makes it the best supported. It is therefore worth spending some time to get to know this operating system.

2. The Pi Models

One of the great benefits of the Raspberry Pi platform has been the continuous development of the product. The Pi Foundation has always worked hard to keep in contact with their customers and this continuous feedback has helped shape the product development. The Pi Foundation have been very discrete when developing their new versions and have the knack of surprising the market with impressive features at a remarkable price point. Perhaps the biggest surprise in recent years was the launch of the Pi Zero at just £4! That model was so cheap, they gave one away free with each copy of one edition of the Magpi magazine. Whilst all this development has been very welcome, the range of models can be a bit daunting to anyone new to the Pi. Whilst the main model B has become increasingly powerful, there are many applications that don't require the features or processing power of the latest model and can work very happily on an older model. In this section, I'll run through the various Pi models, to help you understand their differing capabilities.

Current models

Let's begin with the look at the current Pi range. The flagship model is the Pi-4B, **Fig 17-1**, and this is the model to chose if you're buying for the first time and want to start experimenting with the Pi platform. The Pi-4B uses a quad-core, 1.5GHz 64-bit Cortex-A72 processor combined with Gigabit Ethernet, dual-band Wi-Fi, Bluetooth, two USB 2 ports, two USB 3 ports and a choice of 1GB, 2GB or 4GB of RAM. The Pi-4B also includes dual micro-HDMI ports and can support two 4k displays. This combination makes the Pi-4B a seriously versatile computer for a wide range of applications and a serious rival to budget laptops. One important point while running the Pi-4 is heat management. The additional processing power of the Pi-4 increases the heat output and that needs to be managed to avoid the processor throttling-back due to overheating. When using the Pi-4 uncased on the bench you shouldn't need any additional cooling because the new metal packaging of the SoC (System on a Chip) keeps the temperatures at safe levels. However, as soon as you enclose the Pi in a case or add a HAT [1], you will need to increase the cooling capabilities. There are passive heatsinks available, but they are of limited value as they rely on a good airflow for their cooling. By far the best solution is forced air cooling and the 3rd party suppliers will no doubt develop all manner of ingenious devices. One of the early solutions, that I've found very helpful, is the Pimoroni Fan SHIM, **Fig 17-2**. This comprises a compact 30mm fan that's mounted on a tiny PCB and is

Fig 17-1: Pi-4B.

Fig 17-2: Pi-4B fitted with a Fan SHIM.

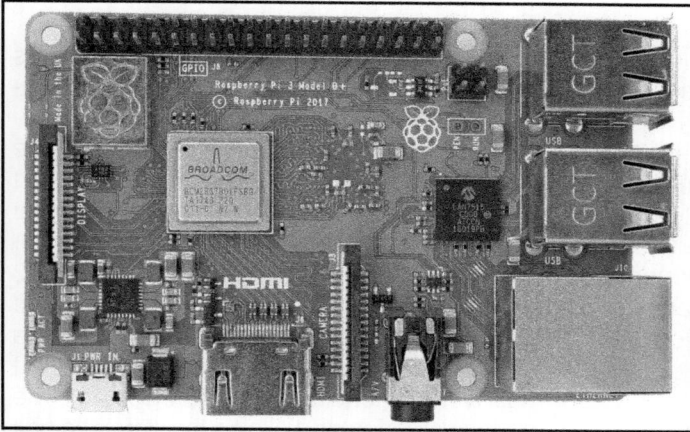

Fig 17-3:The previous flagship Pi-3B+.

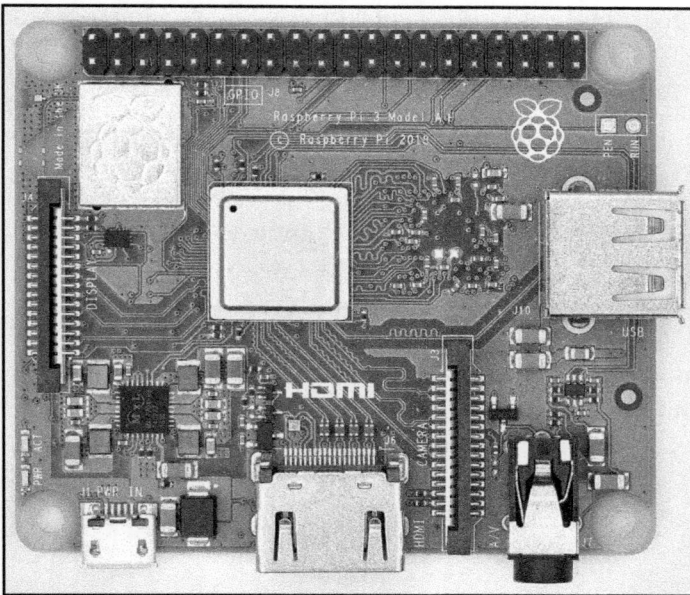

Fig 17-4: Pi-3A+.

Fig 17-5: Pi Zero-W.

a friction fit over the GPIO (general purpose input output) pins. In my tests, the Fan SHIM has dropped the Pi-4 processor's highest temperature by 30°C! In addition to providing excellent cooling, the Fan-Shim leaves all the GPIO pins available for reuse.

For those applications that don't need the power of the 4B, the previous flagship Pi-3B+, **Fig 17-3** is worth considering as the price is likely to drop. For applications that don't require Ethernet and multiple USB ports, the Pi-3A+, **Fig 17-4**, makes a very attractive proposition as it has the same processor and wireless capabilities as the Pi-3B+ but in a Pi-3A+ smaller package that costs around two-thirds of the price of the Pi-3B+.

For even smaller projects, that don't need multiple USB ports and a wired Ethernet, the Pi-Zero range has a lot going for it. The standard Pi-Zero is unbeatable at its super-low price (about £4 at the time of writing). However, if you need access to Wi-Fi or Bluetooth, the Pi-Zero-W, **Fig 17-5**, is often a better answer. In addition to their very low cost and small size, the Pi-Zero models also boast lower power consumption, making them more suitable for battery powered applications.

Model comparison

I've split my comparisons into separate tables showing connectivity, processing power and power consumption to make the information more easily readable.

Pi connectivity

Table 17-1 shows the connectivity comparison between all the Pi models, except for the Compute module. I've deliberately omitted that module because it's primarily intended for industrial use by Original Equipment Manufacturers (OEMs) and is not cost-effective for small projects. Although we are currently on version4 of the Pi, the original generation boards did not have a numerical marking, so if your board just says Pi Model A or B, you can safely assume its version 1.

Raspberry Pi processor table

This section looks at the differing processing capabilities of the Pi models. Like many modern Single Board Computers (SBC), most of the hard work is done using a SoC (System on a Chip) integrated circuit. In these ICs the processor, graphics processor and other key components are combined into a single chip. As these components are usually hard-wired together, combining them in one chip reduces the circuit board complexity and keeps the overall cost down.

You will see from **Table 17-2**, that the Pi's processor has gone through several iterations. However, the graphics processor element has remained stable with a VideoCore IV providing full HD quality graphics support for all models except thePi-4. This current model uses the VideoCore VI processor for dual 4k resolution video outputs. The Pi-3 was the first to use the Cortex A53 64-bit processor, whilst the Pi-3B+ and A+ use a slightly different version of the processor that has a metal enclosure. This improves the heat dissipation and allows the default clock speed to be increased to 1.4GHz, without any additional cooling. The Pi-4 processor looks identical to the Pi-3 but is in fact a completely new and much more powerful device.

Raspberry Pi power consumption table

The power consumption will vary for different applications, so in **Table 2-3**, I've attempted to capture the consumption in a few common

Pi Model	GPIO Pins	Ethernet	Wi-Fi	Bluetooth	USB2	USB3	Serial Camera	Serial Display
Pi-A	26	N	N	N	1	N	Y	Y
Pi-B	26	Y	N	N	2	N	Y	Y
Pi-A+	40	N	N	N	4	N	Y	Y
Pi-B+	40	Y	N	N	4	N	Y	Y
Pi-2B	40	Y	N	N	4	N	Y	Y
Pi-3B	40	Y	802.11n	4.1, LE	4	N	Y	Y
Pi-3B+	40	Y [1]	2.4+5GHz 802.11 b/g/n/ac	4.2, BLE	4	N	Y	Y
Pi-3A+	40	N	2.4+5GHz 802.11 b/g/n/ac	4.2, BLE	1	N	Y	Y
Pi-4B	40	Gigabit	2.4+5GHz 802.11 b/g/n/ac	5.0, BLE	2	2	Y	Y
Pi-Zero	40	N	N	N	1	N	Y [2]	N
Pi-Zero-W	40	N	802.11n	4.1	1	N	Y [2]	N

Note 1: The Pi-3B+ provides fast Ethernet using the USB controller and is faster than previous models but doesn't provide true Gigabit speeds.
Note 2: The serial camera port is only fitted to later Pi-Zeros and uses a narrower gauge socket.

Table 17-1: Raspberry Pi connectivity chart.

Pi Model	SoC Device	Processor	Cores	CPU Clock	Video	RAM	SD/MMC
Pi-A	BCM2835	ARM 1176JZF-S	1	700MHz	1 x HD HDMI	256MB	SD
Pi-B	BCM2835	ARM 1176JZF-S	1	700MHz	1 x HD HDMI	512MB	SD
Pi-A+	BCM2835	ARM 1176JZF-S	1	700MHz	1 x HD HDMI	512MB	microSD
Pi-B+	BCM2835	ARM 1176JZF-S	1	700MHz	1 x HD HDMI	512MB	microSD
Pi-2B	BCM2836	Cortex-A7	4	900MHz	1 x HD HDMI	1GB	microSD
Pi-3B	BCM2837	Cortex A53 64-bit	4	1.2GHz	1 x HD HDMI	1GB DDR2	microSD
Pi-3B+	BCM2837B0	Cortex A53 64-bit	4	1.4GHz	1 x HD HDMI	1GB DDR2	microSD
Pi-3A+	BCM2837B0	Cortex A53 64-bit	4	1.4GHz	1 x HD HDMI	512MB DDR2	microSD
Pi-4B	BCM2711B0	Cortex A72 64-bit	4	1.5GHz	2 x 4k HDMI	1, 2 or 4GB	microSD
Pi-Zero	BCM2835	ARM 1176JZF-S	1	1GHz	1 x HD mini-HDMI	512MB	microSD
Pi-Zero-W	BCM2835	ARM 1176JZF-S	1	1GHz	1 x HD Mini-HDMI	512MB	microSD

Table 17-2: Pi processing power comparison chart.

Pi Model	Board only *Note 1*	HDMI and Ethernet	HDMI and Wi-Fi	USB Max current draw
Pi-A	200mA	n/a	n/a	500mA
Pi-B	500mA	n/a	n/a	500mA
Pi-A+	180mA	n/a	n/a	500mA
Pi-B+	330mA	n/a	n/a	600mA/1.2A
Pi-2B	180mA*	290mA*	n/a	600mA/1.2A
Pi -B	260mA*	380mA*	360mA*	1.2A
Pi-3B+	390mA*	490mA*	520mA*	1.2A
Pi 3 Model A+	190mA*	n/a	290mA	*Note 2*
Pi-4B	560mA*	680mA*	710mA*	1.2A
Pi-Zero	150mA*	n/a	n/a	*Note 2*
Pi-Zero-W	160mA*	n/a	180mA*	*Note 2*

Note 1: Figures with * are measured, the rest are from the Raspberry Pi Foundation website [2].

Note 2: The maximum for these models is only limited by the external PSU and the micro USB connector.

Table 17-3: Pi power consumption.

scenarios. The results are based on measuring the total consumption of real Raspberry Pi units where the current measured was the total drawn from the USB port connected power supply.

Under the hood changes

In addition to the headline changes between the Pi models, there have been a few technical changes that are worth noting. The first relates to the audio output. The Pi has always relied on HDMI for its main audio, but also has a simple Digital to Analogue Converter (DAC) that uses a Pulse Width Modulation (PWM) line from the processor, combined with a low-pass filter to provide an audio output. In the early boards, this suffered significant degradation due to power supply noise. However, from the introduction of the Pi-3, the audio has been improved with the addition of a new buffer chip between the processor and the low-pass filter. This has helped to clean-up the audio signal, whilst also increasing the available output power. In the early models, the audio output had its own 3.5mm jack, but the original jack has now been replaced with a new multi-pole 3.5mm jack that carries both the audio and the composite video output.

On-board power regulation has also changed significantly. Whilst the USB power socket still accepts a standard 5V input, most of the Pi components operate at 3.3V and below, so additional regulation has always been necessary. In the early boards, regulators inside the Broadcom SoC were used to supply the processor and GPU, whilst a linear 3.3V regulator did the rest. However, with the introduction of the Pi-2, external regulator chips were introduced for all supplies and the Pi changed to more efficient switch-mode regulators. This change also increased the amount of current that could be drawn from the Pi USB ports. The latest Pi models use a dedicated, 5 output, switch-mode regulator chip (MxL7704) to provide all the on-board supply needs for the Pi. The high efficiency of this regulator has helped to further reduce the overall Pi power consumption.

References

[1] https://www.raspberrypi.org/blog/introducing-raspberry-pi-hats/
[2] https://www.raspberrypi.org

Pi Projects

Mike's book goes through some essential things regarding setting up and installing an operating system onto a Raspberry Pi in the Projects chapter of his book. Instructions for this are easily obtained online from varying sources. Below is one of the projects from his book.

Project 2 - WSPR beacon

Weak Signal Propagation Reporter (WSPR) is a low speed radio data transmission mode developed by Joe Taylor, K1JT to probe radio propagation paths. WSPR signals use a very slow data rate and only require a few Hz of bandwidth. The result is a signal that can be detected under extremely poor propagation conditions. WSPR transmissions carry only basic information, which is normally the originating station's callsign, their Maidenhead location and their transmit power. The data rate is extremely slow with each message taking 2 minutes to transmit. Stations providing WSPR beacon facilities usually comprise a computer running the WSPR software connected to the main amateur radio transceiver and antenna system. Whilst this works very well, it does tie-up the main station. The Raspberry Pi offers an alternative solution thanks to some innovative work by enthusiasts. Pin 4 of the GPIO on the Pi is connected to a Pulse Width Modulation (PWM) line on the Pi processor. This has many conventional uses but several users realised that this line could be used to generate radio frequency signals. Initial work focussed on using Frequency Modulation(FM) of the PWM line to transmit music on the VHF/FM band (illegally). However, it was soon noticed that this feature could be used to generate RF signals on the amateur bands and a WSPR beacon was proposed. As WSPR is such a slow-speed data mode, directly generating the RF signal is well within the capabilities of the Pi.

Fig 17-16: WSPR filter board fitted to Pi Zero.

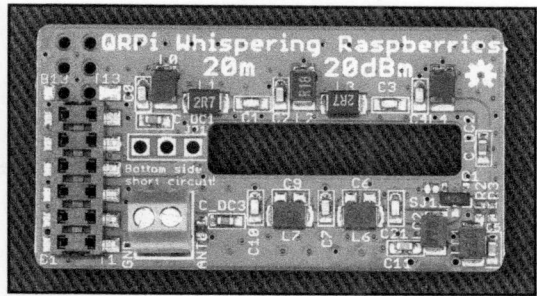

Fig 17-17: The WsprryPi filter board.

What do you need?

1. Raspberry Pi, with an Internet connection (Wi-Fi or LAN)
2. Monitor, Keyboard and mouse (only required for setup)
3. Pi power supply
4. WSPR antenna filter

Filtering and protection:

There are two main problems with using the Pi GPIO pin to generate radio signals. The first is the high harmonic content due to the square waves used on the PWM line. This waveform contains very strong odd-order harmonics and cannot be used without some filtering. The second problem is the risk of damage to the Pi Processor because the GPIO pin is connected directly to the processor chip. As a result, any static build-up on the antenna could easily destroy the Pi processor. Both problems are easily solved by using an external filter board. Whilst there are several good designs on the Web, I opted for a ready-built unit from the Tucson Amateur Packet Radio (TAPR) group. Their filter board uses the design by Zoltan Doczi, HA7DCD and mounts directly on the Pi GPIO pins, **Fig 17-16** and **Fig 17-17**. Rather than just providing a simple low-pass filter, Zoltan uses a bandpass filter that's been optimised to reduce carrier noise. He has also included a sin-gle FET amplifier stage to isolate the antenna from the GPIO pins and added a transient protection diode to discharge any static build-up. The net result is a very compact filter board, with a useful +20dBm (100mW) output that's ready to be directly connected to the antenna.

WSPR Software:

To bring this project to life we need the software to control the PWM pin and generate the WSPR signal. One important point to note here is that the software must be operated with the Pi connected to the Internet. This is necessary because WSPR requires accurate timing to work correctly and the software uses Internet based timeservers to keep the Pi clock accurate. The internet time is also used to calculate the Pi reference clock error to ensure the transmit frequency is correct.

The Wsprry software we need is available for free download from the Github site, and here are the instructions to download, compile and install the software:

1. Open a terminal session (Ctl-Alt-T)
2. Enter: git clone https://github.com/ g4wnc/WsprryPi.git
3. Change to the new directory by entering: cd WsprryPi
4. Enter the following to build the software: make
5. Enter the following to complete the installation: sudo make install

That completes the software installation and you should connect your filter to the Pi and a dummy load to the antenna terminal.

Testing and Operation:

The software has a built-in test-tone facility that is very helpful when checking the basic function of the board. Here are the steps to generate a test-tone:

1. Open a terminal session (Ctl-Alt-T)
2. Enter: sudo ./wspr --test-tone 14.097e6
3. To stop the transmission enter: Ctrl-C

This will begin transmission of a single frequency of 14.097MHz. You can now use an oscilloscope, RF voltmeter or another receiver to check the signal quality. If all is well, you can start a WSPR transmission with the following:

sudo ./wspr --repeat --offset --self-calibration G4WNC IO90 20 20m

In this example, you need to replace G4WNC with your callsign and IO90 with the first four characters of your Maidenhead locator. The penultimate entry (20) is the output power in dBm whilst the final item (20m) is where you specify the band to be used.

When you're happy that the transmitter is working OK, you can connect the antenna and start transmitting. You will find more detailed instructions for using the software on the Github site at: https:// github.com/g4wnc/WsprryPi

When your Pi-transmitter has been running for a while, you should head over to the WSPR website (wsprnet.org) to see if anyone has heard you.

When you reach the site, select the Database tab. This will take you to a screen where you can query the database to see if your transmission has been heard, **Fig 17-18**. Here's a guide to setting up the query:

1. Select the band you're using
2. In the Call box enter your callsign
3. Leave the other settings at the default and click the Update button.

This will return a list of all the stations that have heard your call along with useful information such as the signal to noise ratio (SNR) of your signal and distance to the receiving station.

Fig 17-18: WSPRnet database search page.

Summary

The WsprryPi software we've used for this project makes very light demands on the Pi and so is a good way of utilising one of the less powerful or older Pi models, such as original Pi-B or a Pi-Zero-W. You can also continue to use the Pi to run other software whilst running WsprryPi.

Other Single Board Computers

The Raspberry Pi's unrivalled success is testament to the excellent choices made by the Raspberry Pi team in the overall design of the Pi. Not surprisingly, many others have attempted to benefit from the Pi's popularity by producing a range of look-alike boards. At the heart of the Pi and its competitors is the use of a System on a Chip (SoC) device. This can take many forms but, in the case of Single Board Computers (SBC), the SoC is a single integrated circuit package that typically contains several processor cores, a Graphics Processing Unit (GPU) and other supporting elements, **Fig 17-19**. Combining these functions in a single chip reduces the physical space requirements and helps simplify the printed circuit board (PCB) layout. This is because the complex 32 or 64-line bus interconnections between processors, GPU, etc. are made inside the SoC. The driver for the development of these fast and highly integrated devices has been the smartphone market. Smartphones sell globally in huge quantities and many users expect to upgrade their devices every year or two. This

Fig 17-19: Pi System a Chip SoC.

huge demand funds the high development costs and we get to enjoy the spoils through access to a source of constantly improving SoCs. The Pi has enjoyed very close relations with Broadcom since its inception and all the Pi models use a Broadcom SoC at their core. The other SBCs that I've covered here use either the Rockchip or Allwinner SoC devices.

These Pi competitors usually have a few performance tweaks so they can be advertised as offering an advantage over the original Pi. You should be very cautious when reading these comparisons as they are usually skewed to make the most of the SBC they're promoting.

You also need on to check the Pi model used for the comparison. I've seen examples where the comparison is made against a Model B Pi. However, this often turns out to be the original Model B with its single-core processor running at 900MHz. That's a long way from the current Pi-3 Model B+ with its 64-bit quad-core processor running at 1.4GHz!

The most common enhancements are to add full Gigabit Ethernet and USB 2 ports. This addresses a known bottleneck in the Pi and is relatively easy to achieve with readily available single-chip solutions. However, these enhancements increase both the cost and power consumption of the board. Whist Gigabit Ethernet or USB 2 ports can be useful for some applications, this needs to be matched with a faster processor to handle the increased data load. In addition to a few performance and connectivity enhancements, many of the other SBCs support running the Android operating system, if that is important for your application.

One of the Pi features that has worked particularly well is the stand-

ardised, General Purpose Input Output (GPIO) connector. Even though this was increased from 26-pins to 40-pins early in the Pi's development, the first 26 pins retained their original functionality. This consistency is important as it gives 3rd party manufacturers the confidence to develop their accessories in the knowledge that they will continue to be useful across a wide range of Pi models. The result is a very vibrant and competitive market for Pi accessories. Many of the competing SBCs retain the Pi GPIO layout and functionality so they too can benefit from the established accessory market.

SBC Operating Systems

When considering moving to a different SBC you need to check the availability of Operating Systems (OSes) because standard Raspbian will only run on Raspberry Pi hardware. Without a stable and well-developed OS, the speed and power of the new SBC may well be unusable. The best way to check-out SBC OSes is to read online reviews and check the forums for other people's experiences.

ASUS Tinker Board:

This was launched back in 2017 by this major PC hardware manufacturer, **Fig 17-20**. The Tinker Board is well built and matches the physical features of the Pi very closely. So close, in fact, that it fits the Official Pi cases, **Fig 17-21**. Processing power comes from a Rockchip ARM Quad-core CPU that operates at speeds up to 1.8GHz. This is supported

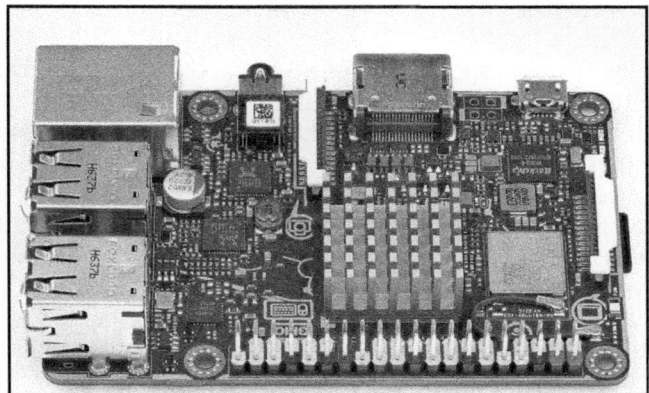

Fig 17-20: ASUS Tinker Board.

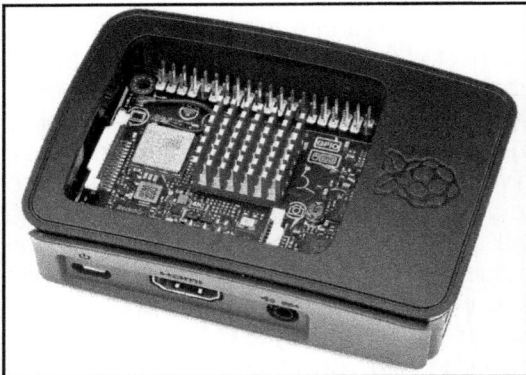

Fig 17-21: Tinker Board in a Pi case.

with 2GB of LPDDR3 memory. As in the Pi, the Tinker Board has a dedicated Graphics Processing Unit (GPU) and can handle UHD video at 30FPS. Another welcome addition to the Tinker Board is the integrated 192ksps/24-bit audio CODEC so there is no need to use an external soundcard. The four USB 2.0 ports have a dedicated controller that is independent of the Ethernet controller, thus negating that potential bottleneck. The network port also supports full Gigabit speeds. Full Wi-Fi and Bluetooth LE is provided but it doesn't have support for the higher speed 5GHz Wi-Fi as provided in the Pi 3B+ and A+ boards. As the Tinker Board uses a Rockchip processor, it won't run Pi Raspbian. However, ASUS do provide a custom Linux image for free download. This image was rather fragile when first launched and receivedlots of criticism, but it has improved over time. An excellent alternative is the ARMbian variant of Linux that's available for many of SBC models including the Tinker Board. I've used ARMbian with the Tinker Board and found it to be very stable. When launched, the Tinker Board was very competitively priced, but the price difference has gradually increased. At the time of writing, the standard Tinker Board was available for £59.99 whereas the Pi-3B+ was £32.81, from the same supplier. ASUS have recently launched an updated version in the Tinker Board S. The main processing power remains unchanged, but it includes a few operating refinements and an onboard

16GB eMMC storage device for faster booting and file access. The new model also carries a significant price increase to £99.

ODROID:
These SBC's are produced by the South Korean company Hardkernel and include a range of powerful designs. The only models that retain significant Pi compatibility are the C1+ and the C2. The C1+ uses a quad-core, 32-bit ARM CPU running at 1.5GHz and supported by 1GB of DDR3 SDRAM. The onboard video processor is a 2-core ARM Mali unit that is capable of handling 4k 60FPS video, which may be important for some, especially if using the ODROID as the heart of a media-centre. Pricing for the C1+ is also very competitive at £40.90. The more powerful C2, features an all-round performance boost with a 64-bit ARM quad-core CPU running at 1.5GHz with 2GB of DDR3 SDRAM. The Mali GPU capacity is also increased to include 3-pixel and 2-shader cores and supports 4k 60fps video. The more powerful C2 comes at a premium and the current cost is £71.20.

Banana Pi Range:
There are a wide range of boards available from this Chinese manufacturer but here I'll look at three that are closest to the Pi format. The Banana Pi M2 Berry follows the Pi model B format closely, but the significant addition is a SATA port. This makes it particularly suitable for use as a media server or in other applications where fast hard disk access is desirable. The M2 Berry also includes Gigabit Ethernet, Bluetooth and Wi-Fi but not the faster 5GHz Wi-Fi, as available in the Pi-3B+. The M2 processor SoC is an Allwinner A7 V40 unit. If you need more power, the Banana Pi M3 steps up to an Octa-core processor running at 1.8GHz and includes a SATA port for fast disk access. Price for the Banana Pi M3 is around £76 but I found it difficult to find stocks.

The final model I mention here is the Banana Pi Zero that uses the Pi Zero format but includes a quad-core A7 processor along with Bluetooth and Wi-Fi, making it a particularly powerful miniature SBC. However, it was impossible to find a dealer with stocks at the time of writing, but I've included it here for completeness.

Nano Pi:

This series of SBCs is from Chinese manufacturer FriendlyElec. However, none of the models use the Pi physical format, though a few include a Pi compatible GPIO connector. The A64 features a quad-core 1.152GHz CPU and a Mali GPU supported by 1GB DDR3 RAM. In addition to standard USB 2.0, the A64 includes a full Gigabit Ethernet port and a Pi compatible 40-pin GPIO connector. The A64 is available direct from China for approximately £20 plus carriage and import taxes. The NanoPi M1 is a low power option that has a quad-core 1.2GHz processor, 10/100 Ethernet, 3 x USB2.0 ports and onboard audio. The NanoPi M1 is available direct from China at £15 plus carriage and import taxes.

Orange Pi:

There are a large range of board variants available from this Chinese manufacturer, but I'll just look at a couple here. The Orange Pi 3 uses a quad-core 64-bit, 1.8GHz ARM Cortex A53 based SoC that includes a Mali-T720 GPU. The processor is supported with 1GB DDR3 RAM and the Ethernet is standard 10/100Mbs. USB connectivity includes 4 x USB 2.0 ports plus a USB On The Go (OTG) port. This allows the Orange Pi 3 to look like a USB device when plugged into the USB port of another computer and loaded with the appropriate software. The Orange Pi 3 also features an audio CODEC and accepts video.input at a maximum rate of 1080p and 30fps. At the time of writing the Orange Pi 3 was only available direct from China where the price was $39.90US. The second device I'll cover here is the Orange Pi PC Plus that is another Pi look-alike based on an Allwinner H3 quad-core Cortex A7 processor with HDMI

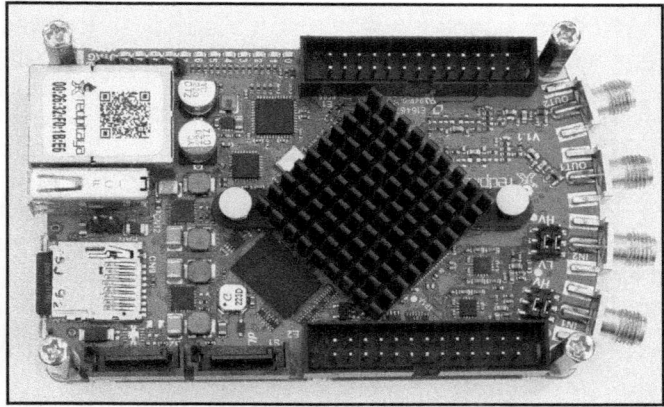

Fig 17-22: Red Pitaya.

4k video support and 1GB DDR3 RAM that's shared with the GPU. The Orange Pi PC Plus also features 8GB of eMMC flash that can be used to store the operating system. This enables faster boot times along with rapid file access. The Orange Pi PC is again only available direct from China at $41.90US.

Red Pitaya:

This is a bit of an odd-ball to include here because it has no obvious relationship with a Pi but is a form of SBC and has many features that make it attractive for amateur radio use, **Fig 17-22**. I have in the past included material on the Red Pitaya in my radio PowerPoint presentations. The Red Pitaya comprises three significant main components:

Dual Analogue to Digital Converters (ADC)

These can read analogue DC through to HF signals and digitise them ready for processing in an SDR receiver. They can also be used to digitise signals for processing with digital oscilloscope or spectrum analyser software. In effect, with the addition of the right software, the Red Pitaya can become a powerful multifunction bench instrument.

Dual Digital to Analogue Converters (DAC)

The opposite of an ADC, these devices convert a digital input into its equivalent analogue output. That enables operation as a signal generator or an HF transmitter.

Field Programmable Gate Array (FPGA)

This is a programmable logic device that can be configured to perform a wide range of functions. One of its major attractions is the ability to parallel process high speed data, thus making it ideal for SDR receivers and test instrument emulation.

The original Red Pitaya featured dual 14-bit ADCs with a maximum sample rate of 125MSPS, when combined with its original price point of around £230 this made it an attractive proposition for the heart of an Amateur Radio direct sampling SDR transceiver for the HF through to 50MHz bands.

Whilst the Red Pitaya has tremendous potential, realising that potential requires one of two options to be available:

1   You need to have the time and capability to master FPGA and SDR programming techniques. This is not a trivial task and FPGA programming in particular is very specialised.

OR

2   You need to locate a source of pre-written software/firmware for the Red Pitaya that will meet your requirements.

The Red Pitaya does have some free, pre-written, test equipment, software available but these have limited capabilities, and, in my experience, many applications have poor user interfaces and are prone to crashing.

From an amateur radio perspective, there is a single engineer that has done some excellent work with the Red Pitaya. Pavel Demin is a young software engineer who has developed code for the Red Pitaya to emulate an HPSDR (High Performance SDR) Hermes transceiver. That means it can be accessed via the Ethernet and operated by any software that's designed for use with the popular Hermes boards. The emulation works very well, and Pavel has also developed a multi-band WSPR transceiver, and an excellent HF Vector Network Analyser (VNA), using the Red Pitaya.

However, to make a practical amateur radio transmitter, you still need to provide a PA stage, adequate, switchable filtering and transmit/receive switching.

In addition to the original 14-bit model, Red Pitaya have launched a cheaper 12-bit model. This has the same sample rates as it's bigger sibling but is restricted to 12-bit samples. The latest model to be announced features 16-bit sampling and a modified sample rate of 122.8MHz, which matches that of the Hermes SDR boards. As you might expect, this is going to be much more expensive and the recommended price is in the region of £510 plus shipping, so probably close to £550. You would also need to add a PA stage, filtering and transmit/ receive switching to form a complete transceiver.

When looking at SDR options with boards such the Red Pitaya, there's a temptation to compare them with popular amateur transceivers and see them as offering a cheap solution. However, a few words of caution are appropriate. The Red Pitaya was designed to be used as a development board in an educational environment. As a result, it is not intended for continuous operation nor for operating in harsh environments. If the board fails you are on your own, as the Red Pitaya team don't offer a repair service. Also, because the board could have been connected to many different devices, it is virtually impossible to prove that a failure was due to a manufacturing fault, as opposed to some external factor. From personal experience, I had a power regulator failure after about two years and Red Pitaya offered no support at all. If you compare a home-built transceiver based on the new 16-bit Red Pitaya with the popular Elad FDM-DUO, it's very hard to make a case for the Red Pitaya. Once you've added the PA, filters, antenna switching, power supplies and enclosures, the cost will be close to £600 or £700, whereas the Elad can be bought with full dealer back up and repair service for about £900!

By far the best role for a board such as the Red Pitaya is as an educational tool to help you develop your own programming skills.

18
Mobile Phones and Amateur Radio

by Lorna Smart, 2E0POI

One of the hurdles that people can face when getting into amateur radio is cost. The cost for equipment such as the radio and antenna can soon mount up. Adding the cost of a computer be it a laptop or desktop can tip the balance. Thankfully the mobile phones are a more affordable option as even a fairly cheap mobile phone can run a lot of mobile applications.

A search on the internet will soon show you how well technology has embraced amateur radio. Like the rest of the technological world, this is an area that is moving at a rapid rate. Not only do apps have to keep up with the ever changing needs of the consumer but they must also keep up with changes to the platforms that they are on and the operating systems that they work on.

The intention of this chapter is to give you an idea on what is out there in terms of amateur radio apps. I have split them into three categories: operation, education and listening. There are apps available that do not fall into these categories but I felt that these were the most common uses that you may want an app for. I have included both free and paid for apps. All the images are from the Android version of the app.

I will be focusing on phones with Android or Apple (iOS) operating systems as these are most common. There will be some crossover as some apps can be used on multiple operating systems and devices. Where this is the case, I will advise. You may also fin that some of these apps will work on tablets as well as mobiles.

Where possible, I have also installed and tried these apps. Unless stated other wise, the Android apps can be obtained from Google Play and the Apple (iOS) from the Apple App Store.

Operation

EchoLink is a free app for iOS and Android developed by Synergenics LLC. It allows you to connect to and talk to any repeater that is connected to the EchoLink network via WiFi or a mobile phone data connection. You will be asked for your callsign then asked to provide an email address and password. Once this has been done a verification email is sent asking you to provide a copy of your licence. Two options are provided for this: to scan a copy and send it online or to fax a copy to them.

Pocket RxTx Free is a free app for Android (contains ads, there is also a pro version) developed by Dan Toma YO3GGX. It acts as a pocket transceiver, allowing the user to remotely control various radio services through their mobile phone (also works on a tablet). It also provides access to some of the WebSDR servers hosted around the globe and information on changes in the world of amateur radio. This is a more complex app but works well once you get the hang of it.

DroidSSTV – SSTV for Ham Radio is an app for Android costing £6.49 which was developed by Wolphi LLC. With this app you can transmit and receive SSTV on your mobile or tablet. Your phone or tablet can either be connected to your HF radio or sat next to the speakers. Once you have tuned the radio into a SSTV frequency and watch the picture on the mobile or tablet screen. At present it supports ham radio modes Scottie 1, Scottie 2, Scottie DX, Martin 1, and Martin 2.

Fig 18.1: QSO screen.

Education

UK Amateur (Ham) Radio Tests is a free app for Android (with in-app purchases and ads) which was developed by JG Services. It provides mock tests for the foundation, intermediate and advanced radio exams. Like a lot of free apps, it does run adverts alongside but they are not too intrusive. It is straightforward enough to use and provides various options regarding the amount of questions so you can fit in some perhaps during your tea break at work.

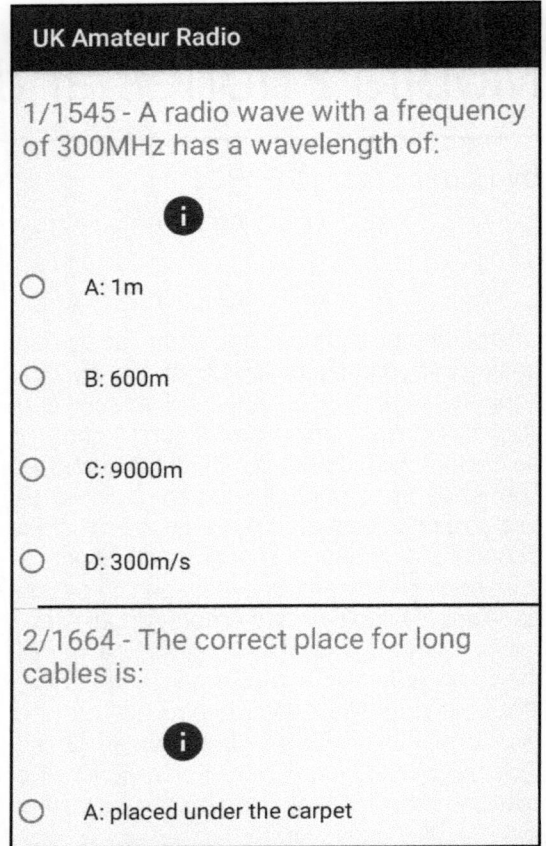

Fig 18.2: One of the many questions on UK Amateur (Ham) Radio Tests.

Ham Mocks(UK) is an app for iOS. It costs $1.99 and was created by John Burns. It is an emulator of the foundation exam. It stores your previous 10 attempts to allow the user to focus on the areas that they are having issues with. There is an extensive bank of questions that follow the RSGB syllabus. It can be used on and offline.

Morse Mania is a free app for Android (with in-app purchases) developed by Dong Digital which mixes education and fun. Learning is often easier when you are having fun. It is a game with the intention of helping the player to learn Morse Code. There are a good amount of levels to make your way through and it can be played in audio, visual or vibration mode. It is a very straightforward

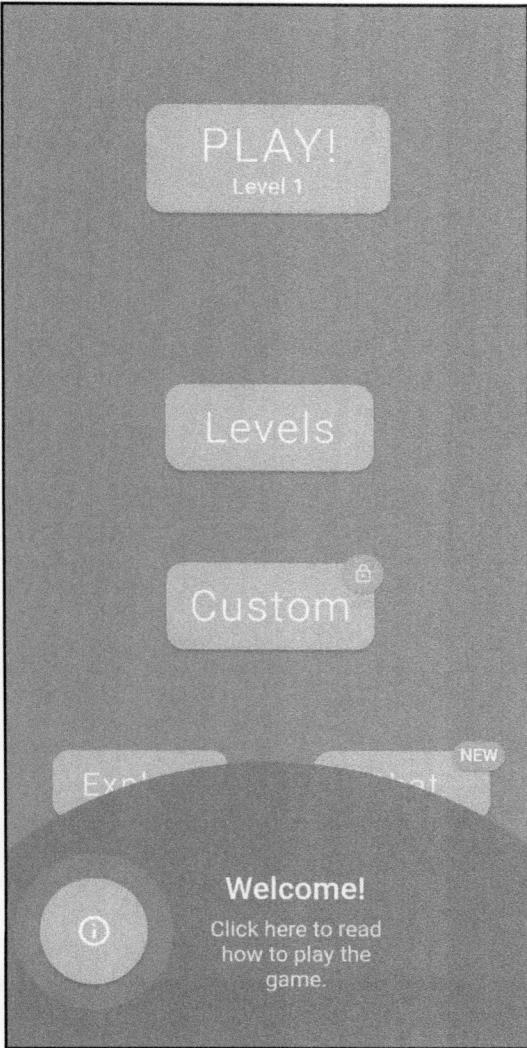

Fig 18.3: Morse Mania menu.

Fig 18.4: Level 1 in audio mode.

set up where you press the letter, number etc that corresponds to the Morse Code that represents it.

Information

Ham Callsigns is a free callsign lookup application available for iOS and Andriod (with in-app purchases). It was created by Robert Allen, N5IKD, in collaboration with QRZ.com and was released in 2020. It is available for Apple, IOS and Android phones and requires no subscription. You do not need to create

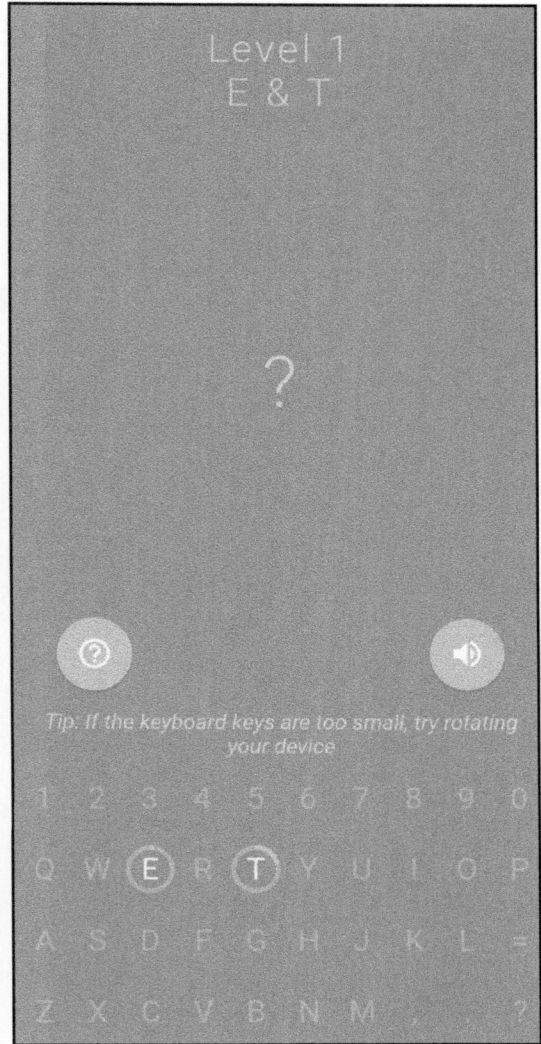

an account or verify your QRZ login details for information such as an address but you do need to do so for email addresses and contact numbers.

Antenna Tool is a free app (contains ads) for iOS and Android developed by Talixa Software & Service, LLC to help with the building of antennas. Using a simple form, the user inputs the frequency and the type of antenna. The app then shows all the calculations that will be required in building the desired antenna.

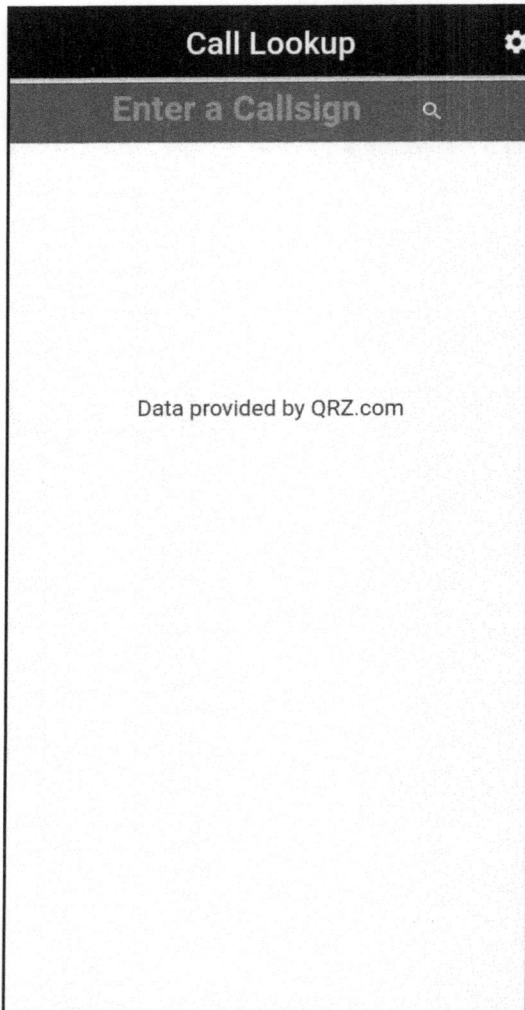

Fig 18.5: Callsign lookup screen.

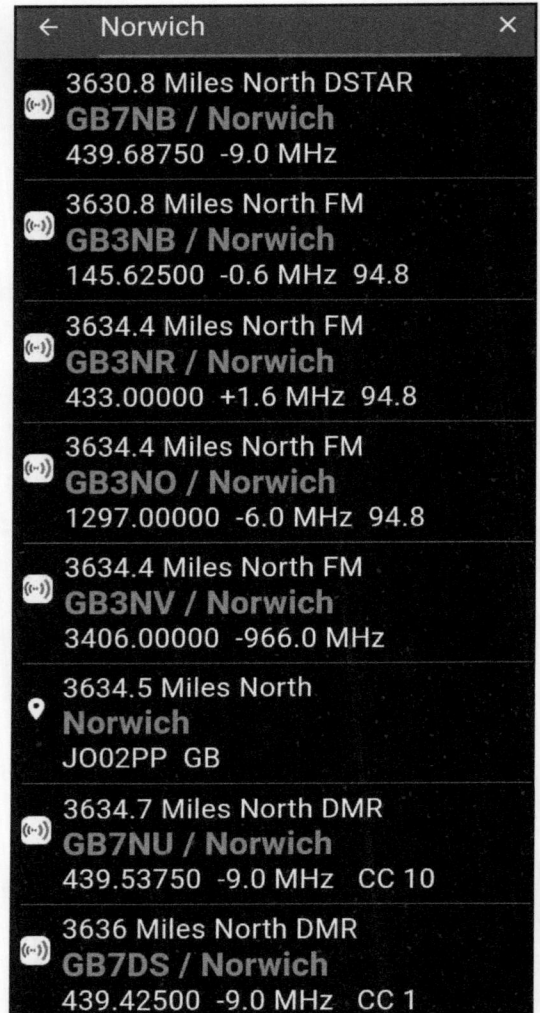

Fig 18.6: Repeater search results.

RepeaterBook is a free local and global app available on both Android and iOS developed by ZBM2 Software. Powered by the database RepeaterBook.com and the software of ZBM2.com, It allows the user to look up repeaters in more than 70 countries worldwide. It supports BlueCAT - FT-857, FT-817, FT897, FT100 and ICOM 7000, 7100, 9100 Bluetooth CAT interfaces and does not require an internet connection using your GPS or network instead to locate repeaters.

There are so many apps out there now for amateur radio. It really is a rich and ever increasing resource pool. It is a good idea to check that the app you are looking at are not country specific (there are quite a lot out there for the US only) if you are looking for something more global or relating to exams (as the requirements differ in other countries) Recently developed apps such as Ham Callsigns show that the mobile phone (and tablet) are taking on a new purpose in becoming part of the arsenal of a amateur radio operator.